T0257524

The Foundations
and Principles of

Biodynamic
Preparations

The Foundations
and Principles of
Biodynamic
Preparations

MANFRED KLETT

Floris
Books

Translated by Bernard Jarman

This book is an extract from *Von der Agrartechnologie zur Landbaukunst* published by Verlag Am Goetheanum, Dornach in 2021
First published in English by Floris Books, Edinburgh in 2023

© 2021 Verlag Am Goetheanum
English version © 2023 Floris Books

Manfred Klett has asserted his right under the
Copyright, Design and Patents Act 1988
to be identified as the Author of this Work

All rights reserved. No part of this publication
may be reproduced without prior permission
of Floris Books, Edinburgh
www.florisbooks.co.uk

 Also available as an eBook

British Library CIP available
ISBN 978-178250-843-4

Contents

Foreword

In June 1924, Rudolf Steiner presented the series of lectures known as the Agriculture Course in which he set out a new way of thinking about the relationship of the earth and the soil to the formative forces of nature. In particular he emphasised how the health of the soil and of the plants depends upon bringing nature into connection with spiritually creative forces that stream in from the cosmos. The practical method Steiner outlined for treating the soil and compost was intended above all to revitalise the soil, which was beginning to deteriorate as a result of modern agriculture and its reliance on chemical fertilisers.

At the heart of Steiner's method are the biodynamic preparations: the two spray preparations and the six compost preparations. Much has been written and spoken about them over the years and still there is always more to learn. Working with the preparations raises questions that demand a great deal of study to answer. Yet this struggle to understand how the preparations work leads to a deeper understanding of life, nature and the human being, and this in turn supports the creative and intuitive capacities needed to cultivate the land. All too often we limit ourselves to the outer appearance of things and processes instead of entering into their wider context, and appreciating the subtle influences streaming in from the wide expanse of the heavens or acknowledging the spirituality that lies at the foundation of our physical existence. These are all key elements for understanding biodynamic agriculture.

In this book Manfred Klett provides a thorough study of the preparations that goes beyond simply setting out how they are made and applied. Instead, he describes in great detail the substances and forces at work in the preparation materials, how they are transformed through the various stages of production, and the effects they have on the soil and compost. As well as making a detailed study of each of the preparation plants, Klett also considers the various animal sheaths used in preparing the preparations, the reasons why Steiner chose them, and how they relate to the organism of the farm.

The role played by human beings is crucial to Steiner's biodynamic method. The core activity of biodynamic farmers is to individualise that part of nature – the farm – which is under their care. This means taming, drawing out its unique qualities and bringing it closer to the human being. Biodynamic practice then becomes a form of artistic activity. Just as the artist uses brushes, paints and canvas to create a dynamic whole work, so does the farmer and gardener work to create this new individual entity, and by applying the preparations invigorate the soil and create new possibilities for development. This far-reaching goal is beautifully summed up in verse by Rudolf Steiner, quoted by Klett:

Unter Schmerzen hat unsere	The earth, our mother,
Muttererde sich verfestigt.	became solid through pain.
Unsere Mission ist es,	It is our mission to bring it
sie wieder zu vergeistigen,	once more to the spirit
indem wir sie durch die	by transforming it
Kraft unserer Hände	through the power of
	our hands
umarbeiten zu einem	into a spirit-filled
geisterfüllten Kunstwerk.	work of art.

Working to understand the preparations and the biodynamic approach in this way leads to a new appreciation for work and research. It is an impulse that arises out of a heartfelt concern for the earth and the sense of responsibility that we bear as stewards of the natural world – something that is all the more important today. The significance of the preparations for the earth and humanity can thus only be grasped on the path of spiritual knowledge.

Klett approaches the subject with a deep appreciation for nature and agriculture, and his thorough knowledge of anthroposophy is evident throughout. This book will provide fertile inspiration for anyone wishing to study the biodynamic preparations in greater depth.

Bernard Jarman
November 2022

The Spray
Preparations

1: Horn Manure (500)

The purpose of the horn manure and horn silica spray preparations is to improve the vitality of the soil, encourage healthy root growth and enhance ripening. When making horn manure preparation, cow horns are filled with manure, then buried in the earth over winter. This exposes the formless manure to the crystallising forces of the solid and liquid elements. The horns are dug up in springtime. The transformed manure is diluted in water through a stirring process, and is then sprayed on the soil towards evening, just before sowing or planting, or after mowing and grazing.

Horn manure preparation

During the summer the spirit of the earth breathes itself out and its covering of sprouting plants becomes an image of the forces living in the light, warmth and air of their surroundings. This vegetative growth provides the food for both humans and animals. Ruminant animals, and cattle in particular, take up this growth from the meadows and pastures and from the crops planted as forage. By chewing the cud and through their digestive processes, this forage material is raised to the level of soul experience. In breaking down the material substances of the plant, the cow is able to 'taste' the cosmic forces they contain. Through this activity of tasting the fodder as it is being digested, the cow experiences the unique qualities of the place where the fodder was grown, such as the particular soil and climatic conditions. She perceives these as a mighty network of forces, and through her internally focused condition of wakeful concentration and outwardly dreamy state, her soul or astral body is able to connect itself fully with the etheric body, which in turn reflects back the physical-chemical processes of metabolism. The soul of the cow participates in what is taking place in her body with endless satisfaction. She cannot, however, retain and make use of this living, soul-strengthening quality since she has no separate, personal self. The forces that have been permeated by her soul have therefore to be excreted. This is what gives cow manure its fine and harmonious fertilising power.

During the summer, the sun's warmth and radiance enhances the quality of the substances contained in the fodder. This ripening process reaches a crescendo in autumn when, according to Rudolf Steiner's recommendations, cow dung to be used in the horn manure preparation should be collected from the pastures. This cow dung is a largely amorphous mass of substance that, left to itself, would become part of nature's process of forming humus. However, we then intervene and replace the normal sequence of events with the following procedure.

The first step in making the preparation is to fill cow horns with manure. These should be horns obtained from our own

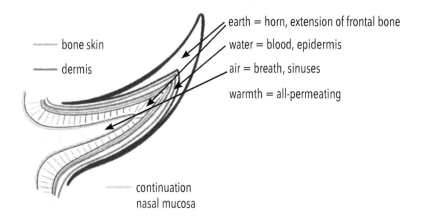

earth = horn, extension of frontal bone

water = blood, epidermis

air = breath, sinuses

warmth = all-permeating

bone skin

dermis

continuation
nasal mucosa

Diagram 1.1: The physical structure of the cow horn and the four elements work together to create an internally focused sense organ.

herd if possible. Bull's horns are unsuitable for this. In its form the cow horn is the opposite of manure. Situated on the head, it is the focal point of the animal's nerve-sense system and is a highly dense formation of ossified skin. The horn envelops a bony core that grows out horizontally from the frontal bone and its air-filled spaces link up with the cow's sinus system. This bony core is covered with a delicate network of arteries that supply blood to the leathery membranes and the growing horn that encases the core. The four physical states of solid, liquid, gas and warmth are concentrated within the pulsating life of the horn, and these form a kind of sense organ for perceiving the internal rather than the external world – a sense organ that serves to hold back and intensify (see Diagram 1.1). What has been streaming out in the bloodstream towards the head is directed back into the cow's metabolism by the horn's dense and strongly formed casing. The cow needs this reflecting-back function of the horns so that the network of forces within the metabolism (unconscious imaginations perhaps) can instil the proper fertilising power in the manure.

With the filling of the horns the first step in making the preparation is completed. We can think of this as a nature process turned inside out: what has been excreted now fills an inner space. The horn, which is so dense that it expresses the principle of pure form, becomes a container for formless substance. Through this process we place ourselves between the nerve-sense and metabolic poles of the cow, and create a new relationship between the horns and the dung: what reached a conclusion in two opposite directions within bovine evolution has now been reunited and becomes a starting point for a new direction of development.

To evaluate this procedure and the further stages that follow, we need to direct our attention to three approaches to research. The first is the framework of ideas that supports and guides the activity, the second is the experience of the activity itself and, thirdly, observation of the life of the cow that shows us the properties and functions of horns and cow manure.

Next, we bury the horns in the earth and leave them over winter. During this time, vegetative life withdraws to a seed-like condition, the earth enters a spiritually awake state and its solid and fluid elements are subjected to the strong crystallising forces of the fixed stars. This is another intervention in the processes of nature whereby the amorphous substance of manure, which would normally play a part in the humus forming processes of summer, is now exposed to the crystallising forces

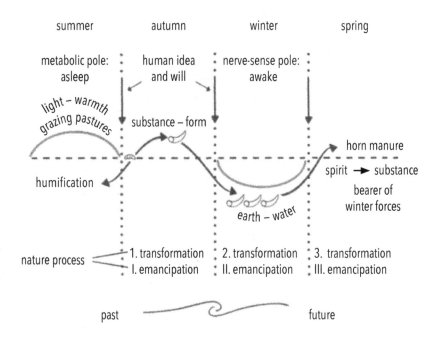

Diagram 1.2: The steps of transformation and emancipation, through the cycle of the year, when preparing horn manure.

of winter. The cow manure becomes a receptive matrix for forces that have streamed into the horn cavity through the physical processes in the solid and liquid elements and that are then retained as concentrated fertilising power in the transformed manure.[1]

Finally, at springtime, we take the horns from the earth and knock out their contents. We now hold in our hands a new substance with new properties. The value and effectiveness of this created substance comes about as a result of it spending the winter in the earth under the influence of the forces active there. It has been 'penetrated and permeated by the spirit'.[2]

The stirring process

A fourth stage is carried out immediately before the preparation is applied. This is the stirring process, in which the preparation is changed from a solid, earth-like condition into a liquid state (see Diagram 1.3).

A small amount of the horn manure is stirred in a rhythmical process in warm water for an hour. A maximum of four horn contents per hectare (about 1½ horn contents per acre) is recommended.[3] The stirring is best carried out using a stirring broom attached to the roof or a strong beam in such a way that it remains mobile. The broom then hangs submerged in a barrel filled with water. We then bring the mass of water slowly into movement by stirring in a circle around the periphery. When a vortex forms, we gradually move the stirring broom towards the centre. At the centre of the vortex there is a tendency towards greater and greater speed (hence the power of suction), while towards the periphery the speed tends towards zero. Between these two poles the differences in speed create circulating layers as the homogenous body of water structures itself in surfaces that run alongside each other from the centre out to the periphery.[4]

When the vortex is fully manifested we disrupt it by holding the stirring broom against it. The structure of the water dissolves, returning to a state of formless chaos, and, for a moment, almost becoming still. Then we stir in the opposite direction to create a new vortex. We thus keep the water in a dynamic rhythm that alternates between the polarities of movement and stillness, homogeneity and the

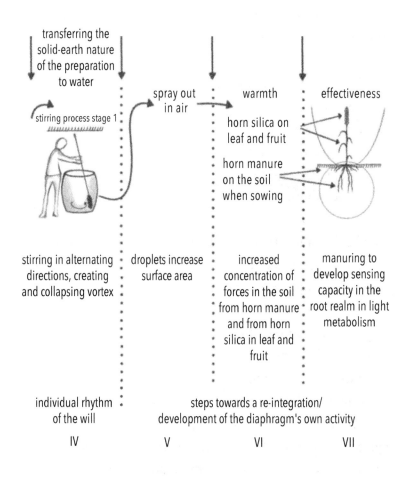

transferring the
solid-earth nature
of the preparation
to water

stirring process stage 1

spray out
in air

warmth

effectiveness

horn silica on
leaf and fruit

horn manure
on the soil
when sowing

stirring in alternating
directions, creating
and collapsing vortex

droplets increase
surface area

increased
concentration of
forces in the soil
from horn manure
and from horn
silica in leaf and
fruit

manuring to
develop sensing
capacity in the
root realm in light
metabolism

individual rhythm
of the will

steps towards a re-integration/
development of the diaphragm's own activity

IV   V   VI   VII

Diagram 1.3: The stirring process, application and effectiveness of the
horn manure and horn silica preparations.

laminar structures of a whirling vortex. These ever dissolving surfaces, formed by the differentiated speed of the moving water, become a matrix for receiving three impulses:

1. The concentrated force potential of the preparation substance.
2. The currently active cosmic constellations.
3. The individual soul-spirit of the people through whom the process has been initiated, guided and carried through.

As has been mentioned, stirring is carried out for one hour. Is this a random period of time, chosen to ensure that the power of the preparation is combined with the water, or might it have some other purpose? The answer is not to be found in the processes of nature but rather in the cosmic rhythms that work through and are retained by the human being. By this is meant the 24-hour rhythm of day and night.

The human 'I', the soul-spirit of the human being, lives within the sun-earth rhythm of sleeping and waking. Macrocosmic rhythms are individualised by the human 'I' or self and impress themselves on the nerve-sense, rhythmic and metabolic-limb systems of the physical body – for instance, the rhythm of the breath or the beating of the heart. The wavelengths of these rhythms are measured in terms of minutes and seconds. In the nerve-sense system they are reduced to a fraction of a second, while at the opposite end of the spectrum, in the metabolic-limb activity, they can stretch to several hours. The metabolic-limb system, however, is home to the will and the activation and deactivation of the will occurs in the time span of one hour. This, along with personal experience, confirms that the one hour of stirring is connected to the rhythm of the will in the human metabolic-limb system, for that is, after all, what sets the stirring process in motion.

Furthermore, living in the unconscious depths of the will is the 'I'. The 'I' is active in the world and, in this case, in the water that is moved through the activity of the will. Everyone familiar with preparation stirring knows that it is possible to maintain continuous activity for an hour by finding a rhythm that reflects their own will activity. Each person imprints their own rhythm on the water, which is dynamised by the laminar flow of the moving vortices. This human 'will rhythm', in contrast to all the rhythms of nature that are of macrocosmic origin, is the first and only rhythm brought by the human being as microcosm to the processes of nature in this fourth stage of preparation.

By delegating this task to a machine the opportunity is lost for entering into the process of stirring in full awareness. The basis for trying to understand it for oneself is likewise lost. The question of why stirring is done for an hour has no meaning if the engagement of the human will is removed. If it were only about achieving an optimal mixing, the whole operation could be completed in a few seconds using a machine. But that is not the issue. Stirring with a machine of any kind, even using the Flowforms developed by John Wilkes,[5] falls under what Rudolf Steiner referred to as the 'slide into substitute methods'.[6] The problem is not about being able to copy hand stirring and the reversing of the vortices using a machine, but that as a result of using a machine, instead of the spirit-led, rhythm-producing individual will, the process is directed from outside using a mechanical beat. Rhythm brings cosmos, earth and the human being together, while a beat brings about separation.

It could be said that an hour spent stirring the preparation is an hour lifted out of the daily work routine. It is an hour worth cultivating. Stirring can be carried out with one other person, with three or more people or by inviting members of the farm-support group to join in. Coming together to engage in a rhythmic process can bring about a relaxed and joyful mood and a feeling of being free in an activity that is self-determined while connecting with the serious task of stirring on deeper levels of the will.

Apart from Rudolf Steiner's comments about how stirring should be carried out from a spiritual scientific standpoint, and the observable facts about forming and breaking up a vortex and so on, our own soul experiences of thinking, feeling and willing can take on particular significance. Observation of ourselves shows that the rhythmical process of stirring is reflected in the way each form of soul activity – thinking, feeling, willing – relates to one another. Before I make the decision to bring the water into movement these three soul forces form a unity in me. I am entirely myself and stand in front of something external. The moment I consciously decide with my thinking to move the stirring broom, the will becomes active. This activity steadily increases, my will drives the water round without pause until I reach my limits, which is when the vortex can be enhanced no further. Meanwhile, my thinking separates itself from the will activity and withdraws to the point where, in complete stillness and with heightened awareness, it can look upon its own activity. My soul experience is divided between stillness of thought and mobility of will. In the tension created by the developing vortices between these two poles, my feeling life can unfold independently of thinking and willing in a pure and spiritually open way. Moments of enhanced presence of mind can then occur. I can experience myself between stillness and movement and of being at one with the outer activity. I am right in it. It is here, lying within this sense-free realm of feeling created by my own activity, that the deepest source of inspiration for crafting true certainty on a chosen path can be found.

If the will no longer has the strength to increase the speed of the water, our thinking intervenes once more and causes the will to collapse the vortex by abruptly changing direction. In that moment of change the water becomes a homogenous mass and the separated soul qualities of thinking, feeling and willing reunite. Out of this oneness comes the decision to form a new vortex in the opposite direction, thereby causing the soul forces to separate once more. In this stirring process the focus of the soul weaves rhythmically between self and world. Through this portal the 'I' manifests itself in the world and within this same rhythm the being of the world arises in the 'I'.

Application

As soon as the stirring process is complete, the horn manure preparation is applied in large droplets, usually in low volumes of 40–60 litres per hectare (4–6 US gallons/acre) or even less (see Diagram 1.3). The horn silica preparation is also applied around the same time (see next chapter).

What was once solid and was then rhythmically dissolved into liquid now opens up into the air and warmth element through the droplets of water. The constellation of forces present at the time of application impresses itself upon each droplet, which opens outwards on all sides to the peripheral forces active in the light element.

Horn manure is applied at the time of sowing on the cultivated soil (ideally at the same time as the sowing), and for a second time when the land is harrowed in spring. On grassland it is applied in the autumn and in the spring after grazing or when the hay has been cut, and in the orchard just before bud burst and after harvest. Late afternoon is the preferred time for stirring and spraying, especially when the sky is overcast.

No hard and fast rules can be given regarding application times. It depends on our own personal relationship to the crops and the place they have in the whole farm. The times emerge intuitively and often quite precisely through an inner connection to the work and by reflecting on our observations and experiences.

2: Horn Silica (501)

For the horn silica preparation, silica or rock quartz is ground to a fine dust. What was crystalline is now reduced to a formless mass. Water is added to create a paste and this is poured into cow horns. The filled horns are then buried in the earth over the summer, exposing them to the forces at work in the airy warmth that permeate the soil at this time. The horns are dug up in autumn. The transformed paste is diluted in water through a stirring process. The potentised liquid is then sprayed on the plants in the early morning.

Horn silica preparation

The first two steps in the process of making the horn silica preparation are the opposite to those for making horn manure (see Diagram 2.1). In contrast to manure, which can be seen as a metabolic product of summer, the crystalline quartz represents the winter activity of the earth. Silica (quartz, rock crystal) is the 'universal sense [organ] within the earthly realm',[1] able to withstand the forces of weathering. Once again, a turning inside out occurs when we pour the quartz paste into the horns:

what is outside becomes inside. The second step is to bury the filled horns in the earth, this time to be left over summer. In this way the winter-related forces of quartz become substances of summer. The amorphous condition of the quartz enables it to become the receptive matrix for the metabolic forces that permeate the air and warmth of summertime. Caught and rayed into the cow horn cavity, they concentrate themselves in the prepared quartz powder. This is able to retain the summer forces in the same way that horn manure holds on to the forces of winter.

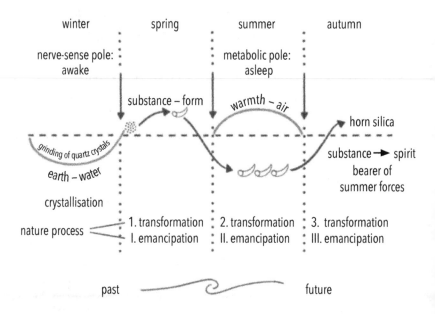

Diagram 2.1: The steps of transformation and emancipation, through the cycle of the year, when preparing horn silica.

We might wonder why we bury the silica horns instead of leaving them above ground where they would be more directly exposed to the warmth, light and air of summer. My impression is that it is not so much about the concentration of forces in the cow horn as they work *above* the earth, but how these same forces become effective *indirectly* through the clay, in so far as they are drawn down by the power of calcium via the roots of plants. When, as a third step, we dig up the horns in autumn, we once again find a new substance with new properties – this is the horn silica. Its nature arises out of the three steps described, which takes it from a mere natural phenomenon to one transformed by human ideas and actions. The finished preparation is shaped by the substances working in the earth during the summer, a concentration of 'material that is woven through by spirit'.[2]

Both horn manure and horn silica are created substances. They are not the result of a natural process that comes to an end and is limited – as is the case with soil, manure, quartz and horn – to what nature can bring about during the course of the seasons. They both arise through having been transformed three times.

Stirring and application

The stirring process for the horn silica preparation is the same as it is for the horn manure. A small amount of the preparation is stirred in warm water for an hour. We begin by stirring on the periphery, before moving gradually towards the centre as a

27

vortex forms. We then collapse the vortex and begin stirring in the opposite direction.

Once the stirring process is complete, the horn silica is applied as a fine mist over the plants. This is done in similarly low volumes as the horn (40–60 litres per hectare or 4–6 US gallons/acre or even less), and is generally applied twice or more during the period of growth and ripening. The specific times are chosen in accordance with the growing cycle of each individual crop. The general rule is that times for spraying the horn silica preparation are chosen to support and harmonise the growth and ripening phases of the particular crop. The time of day is chosen when the earth is breathing out and the sap is rising. The preparation is usually stirred in the early hours of the morning and sprayed on plants still damp with dew. For crops ripening in the vegetative phase, such as root crops, a spray given in late afternoon or evening as the sap is descending is recommended.

How the spray preparations work

On the journey described, both preparations have followed the reverse path to that normally taken in terms of the elements: from the conditions of the solid-earth element they have ascended through water and air to the moment when, as droplets, they arrive as warmth. In a very brief moment all that can be perceived as, or is susceptible to, gravity either gets drawn into the earth or evaporates. By going through this condition of warmth, the pure spiritual force, which has been concentrated in the manure or silica in the cavity of the horn during the preparation procedure, becomes effective.

This four-step journey must be accompanied on an outer and an inner level if we are to truly understand Steiner's description of 'how the dung from the cow horn drives from below upward, while the other draws from above'.[3] What is referred to here are pure force effects in the living realm that have come into being through a process taking place in time and space. They now work into earthly conditions in such a way that, from the moment of germination, through the stages of growth to the fruiting and ripening phase, the plant can fully integrate itself according to its type, the site condition and the vertical sun-earth axis. The quality of being lives in warmth and manifests itself through warmth.

The horn manure preparation becomes effective beneath the earth in the root realm or, as we might say, in the 'head' of the farm organism. It fertilises the processes beneath the earth that may be compared to the nerve-sense processes that occur in the human head. These fertilising forces have been absorbed during the winter when the horns were lying in the earth, during which time activity is naturally most intense inside the earth. With horn manure we have in our hands the 'fertilising power of winter forces', which can be applied not in winter but in the spring, summer and autumn seasons when crops are being sown. The cow manure, with its active metabolism, has been transformed into a fertiliser with an active sensory function that can strengthen the activity of the roots themselves rather than simply the minerals in the earth. It 'educates' the sensing capacity of the roots. According

29

to Steiner, the roots of plants 'can be compared to the organ in human beings that looks at things, our eye, but in human beings it's a weak organ.'[4] The effect of horn manure is confirmed by experimental research showing that roots grow truer to type, are more finely divided, penetrate more deeply into the soil and so gain access to a larger area of ground.

The horn silica preparation becomes effective above ground in the realms of shoots and leaves. Here, in the 'belly' of the farm organism, intense metabolic processes take place through the exchange of light, warmth and air. In this mutual interplay between forces from the periphery on the one hand and the rise and fall of plant sap on the other, the horn silica preparation brings a harmonising effect. In this sense it is a 'fertilising of the plant's metabolism'. This fertilising power has been absorbed while the horns lay beneath the soil during the summer, the time when the summer forces of metabolism are at their most intense. The crystalline quartz with its active sensory quality – a representative of the winter soil – has been transformed by the processes described earlier into a summer fertiliser with an active metabolic function. With its help the effects of the great macrocosmic rhythms of the year's cycle can be individualised in the crops we are growing.

This manifests itself clearly in the appearance of the plant, for instance in the more fully formed leaves, in the leaf metamorphosis and in the composition of substances during the burgeoning growth period, and in the devitalising stages of flowering, fruiting and seed formation. In cooperation with the horn manure preparation, it brings harmony to the processes of fruit formation and, as experimental trials show, ensures physiological stillness in the ripening process. In its consistency, colour, scent and flavour it has qualitative characteristics that are beyond the threshold of mere plant nature and provides quality nutrition for animals and humans.

Horn manure and horn silica require each other. They work in the earth-sun axis that manifests itself in the vertical gestures of root and shoot. They also open up the plant to the forces of the periphery, which the intense branching of the roots and the intricate forms of the leaves demonstrates. With one preparation we are fertilising with winter forces and with the other, summer forces. We fertilise life with what bears life. The preparations are living substances that release forces that make the rhythm-forming earth-cosmos relationship individual to the particular site.

The Compost
Preparations

3: Yarrow (502)

The purpose of the compost preparations is to transmit vitality to the compost and then in turn to the soil and the plants growing in it. The preparations stabilise the soil's nitrogen content, promote microbial diversity, and regulate the decomposition and humus-forming processes. They work with various substances that are essential for healthy plant growth, such as sulphur, potassium, calcium and phosphorous, and bring them into right relation with the growing plants.

The first three preparations work within the earth, in the 'head' of the farm organism or garden, and draw nearer to the surface of the soil, which forms a kind of 'diaphragm' or middle region. The other three preparations work more with the life processes that exist above the soil in the 'metabolic' pole, which relates to the elements of air, warmth and light.

The first compost preparation that Steiner describes is the yarrow preparation. This is connected especially with sulphur and potassium, which work together to form protein and thereby build up the structure of the plant. Sulphur can be seen as the bearer of the creative spiritual principle that comes to expression in the form of the plant, whereas potassium, as a salt, is the bearer of material substance and is representative of what is earthly. Through each stage of development, the potassium is increasingly released from its terrestrial nature, becoming more refined and rarefied as it is dissolved. This process culminates in the flower: the part of the plant that opens up to the influences streaming in from the cosmos. The yarrow preparation seeks to preserve this moment of flowering and the forces that underlie it. It revitalises soil that has been exhausted through over-cultivation.

Yarrow's characteristics

Yarrow is described by Rudolf Steiner as a 'miraculous creation'. He said, 'Yarrow stands out in Nature as though some creator of the plant world had had it before them as a model to show them how to bring sulphur into a right relation to the remaining substances of the plant.'[1] Then, later: 'Yarrow mainly develops its sulphur-force in the potassium-formative process. Hence it has sulphur in the precise proportions which are necessary to assimilate the potassium.'[2] In yarrow the sulphur works on the potassium content so that 'it relates itself rightly, within the organic process, to that which really constitutes the body of the plant.'[3]

Sulphur is the mediator of cosmic-spiritual forces in earthly substances, which organise those substances according to their spiritual archetypes. In the case of yarrow, these forces come to manifestation in the form of the plant. Through the working of sulphur and potassium, two opposite principles, protein is formed in the plant. At each stage of this development, from seed through to the flower, the plant is an image of this polarity, and this is especially true of yarrow.

This can be seen quite objectively when a seed is placed into the earth. The plant's archetype is concentrated in the seed; the cosmic element lives within it as the form of the plant. The seed is then surrounded by humus, the formative principle within the earth, and by the mineral constituents of the soil: silica, calcium, clay and dissolved salts.[4] The seed is thus surrounded by matter whose terrestrial nature diminishes the more strongly its mineral components crystallise and become insoluble in water, as is the case with silica for instance. This is because in crystals the form of the cosmic archetype of the mineral appears, whereas in dissolved substances physical-terrestrial laws prevail.

As the seed swells, water is taken up from the soil along with dissolved salts, and the cosmic form of the plant forms a relationship with earthly substance. What lives as cosmic form within the seed transforms itself, manifesting as earthly form in what we see as the physically perceptible plant. When an earthly substance such as potassium is absorbed, it takes the first step away from its terrestrial nature. In this field of tension between the plant form becoming earthly and the potassium increasingly losing its terrestrial quality, protein is created; cosmic form and earthly substance interpenetrate one another in this fundamental substance of life. For example, when protein is created in the growing shoot, potassium processes rise up and exert pressure from below. This enables the spirit, with sulphur as its mediator, to use the protein to build up the structure of the plant. The forming of plant protein can be seen as the moment when the plant's cosmic form enters the earthly realm and when potassium, as the representative of terrestrial forces, reveals its cosmic-archetypal nature. Hence the more the physical form of yarrow comes to expression the more is potassium released from the constricts of its physical properties.

At every stage of this changing interrelationship of form and substance, of becoming terrestrial or cosmic, the protein has a different form and constitution. In the young plant and towards the root, the structure of the protein is simpler and more related to potassium, to nitrates and free amino acids; whereas in the fully grown plant, towards flowering and the ripened fruits, it grows more complex and is related more to sulphur and to form (see Diagram 3.1).

In the flower, the full expression of what once lived in the seed as cosmic form now appears in earthly form. The being of the plant cannot reveal more of itself than what is manifested in the flower. Potassium on the other hand, as an earthly substance, is furthest from its earthly nature in the flower. From a substance with clearly defined physical properties it has dissolved into the organic process. It has imprinted on this process certain characteristics that allow it, as it ascends from leaf to leaf, to achieve a seed-like state in the flower.

This relationship between the cosmic and terrestrial qualities, expressed in the flower in terms of form and substance, mirrors that of the original seed germinating in the soil, but this time in a polar opposite way. The plant reveals its archetypal nature by dying into form while, at the same time, the earthly substances that permeate it enter into a kind of 'cosmic seed state'. As closed and complete in itself as the flower may appear, it is also open and germ-like at the same time. This condition of openness and devotion towards the cosmos lasts for only a moment. Then comes the impulse that leads to the forming of individual seeds and the fading away of the plant form after flowering, causing it to decay and become humus, the 'universal seed'.

How can this process of seed and humus formation be led beyond the limits set by the laws of nature? How may the moment of flowering be given permanence?

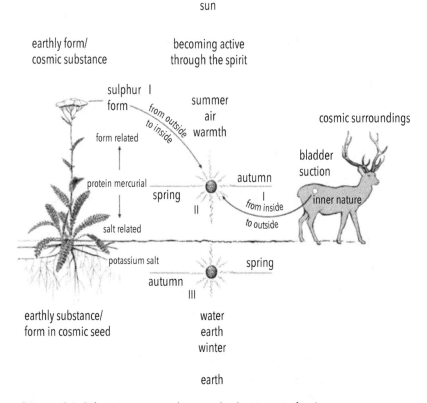

Diagram 3.1: Substance metamorphoses in the three stages of making the yarrow preparation.

Stages of the preparation

Answers to these questions cannot be found merely by observing the yarrow plant. Rather, we must look at the effect that yarrow has on animals and humans. There we find that it has a healing effect on weaknesses in the astral body. This is because animals and humans have an inner life that corresponds to the flowering of plants. When used as a remedy for warm-blooded animals and human beings the flowering process of the yarrow continues beyond the realm of external phenomena and there unfolds its healing properties. This is the effect we are striving to achieve with the yarrow preparation.

According to Rudolf Steiner, the organ process that is best able to conserve what is present in the yarrow is 'the process which takes place between the kidneys and the bladder'.[4] What makes the bladder of the stag eminently suitable for the yarrow preparation as compared with other ruminants, is its two functional aspects.

The first functional aspect is what connects the bladder with the outside world. In the bladder, the metabolising of fluids reaches a conclusion. It draws in, concentrates and then excretes what has been processed by the kidneys. This activity has as its counter pole the outwardly focused nerve-sense activity. In the case of the stag this applies not just to the eyes but also to the antlers. They serve as a kind of feeler or sense organ, and grow out like limb bones above the head and then die off as an external object of the world.

The red deer lives with its proud antlers and ever watchful eyes observing its surroundings in a state of continuous low-level nervousness. The eyes behave in an opposite way to the bladder. While the latter captures material substances from the internal, ensouled world, so in the opposite way the eye captures spiritual impressions in the outer world and concentrates them as perceptions. And just as the bladder opens itself towards the external world through the act of excretion, so the eye transmits the content of its images to the world within.

What the stag experiences as the current cosmic reality influences not just the particular configuration of the bladder but also the makeup of substances in its fine membrane. Through the imprinting of these forces, the round and ball-like bladder almost becomes an 'image of the cosmos'.[5]

The second functional aspect has to do with how the ball-like form of the bladder encloses an inner space. Within this space the ongoing activity of the processes taking place there are preserved.

In the first stage of the preparation, the yarrow flowers are placed, lightly pressed together, into the bladder of a stag so that they are completely surrounded by it (see Diagram 3.1, I).

The flowers that had previously been open to the cosmos now fill an inner space, while the bladder, which was once wholly at the service of inner, organic life, has now become an object of the external world. The astral body of the stag, which formed the bladder, is now replaced by the world astrality of the cosmos, which streams in from the periphery when the preparation is exposed to the outside world. During this first stage of the preparation, we see an initial transformation of the bladder's function taking place: from an organ of metabolism, it becomes a kind of sense organ for sensing the cosmos.

The flowers on the other hand, which are naturally open to the outside world, now find themselves enclosed within an animal organ that is thoroughly permeated by conserving forces.

In the second stage of the preparation, the stag's bladder with its content of blossoms is hung up 'in a place exposed as far as possible to the sunshine'.[6] There it is exposed to the forces of the earth's physical body in the elements of air and warmth and all that works through the vertical axis running between the centre of the earth and the sun. The configuration of substances bestowed on it by the stag's experience of the cosmos has made the bladder membrane receptive to these forces. This signature of the sun-filled space of air and warmth is imparted by the bladder to the flowers contained within it, and the form of the bladder envelope ensures that what has been taken up by the yarrow flowers is retained and preserved.

In the third stage of the preparation the balls of yarrow are buried 'not very deep in the Earth',[7] so that they are exposed to the forces of the earth's physical body in the elements of earth and water and that which is working in a vertical direction. Again, it is the configuration of substances in the bladder, which mediates what is living physically in the earth and the water, and it is the enveloping form of the bladder itself, that gives a lasting quality to the germ-like condition of the substances in the blossoms.

In the second and third stages of preparation, the yarrow flowers are exposed to the forces of the physical earth. This consists of and is built upon the four elements, which in nature are continually mixing and separating out into the air and warmth above and in the fluid and solid elements beneath the earth. The balls of yarrow are submersed as seeds in both of the spatial qualities of 'above' and 'below', and in light and darkness. These spatial qualities are preserved in them, and that which is preserved above permeates that which is below and vice versa.

Besides the physical and spatial context, time also plays a role in this second stage. The balls of yarrow spend summer through to autumn suspended above the earth, exposed to the forces of warmth and air as well as to the light of the sun. During this period, the airy, warm and moist spheres of the earth are permeated by the etheric and astral forces that stream in directly from the sun and also from the planetary surroundings, especially from the inner planets. What is living spiritually in these forces enables the plants to grow and develop in an endless variety of forms. These forces are received by the blossoms inside the sheath. Again, it is the bladder's configuration of substances that imparts to them what lives in the succession of time, and it is the bladder's form that preserves the imprint of each moment.

In the third stage of the preparation, the balls of yarrow lie in the soil from autumn through the winter until spring. The moist earth is permeated by etheric and astral forces from the sun and the planets, but especially the outer planets this time, that work indirectly on plant growth. What lives spiritually in these forces enters into connection with substances representing the mineral, crystalline nature of the earth: quartz (silica), limestone and clay. It works on the plant world indirectly as an upward cosmic stream mediated by the clay, and it manifests itself in the colour of the leaves and flowers and in the refined substances of the ripening fruit. The balls of yarrow lie immersed within these active forces beneath the earth during the winter. Once more it is through a kind of sensory capacity within the substances of the bladder that this cyclic, germinal quality is imparted to the mass of blossom material, and the form of the bladder that preserves this germinal quality.

Summer and winter cannot occur simultaneously in one place, they must follow each other in time. But by being subjected to the forces of light and warmth during the summer and then those of water and earth during the winter, the germinal quality in the flower (mediated and preserved by the stag's bladder) has an entire annual cycle imprinted upon it. All the qualities of the successive seasons of the year permeate the flower substance simultaneously. It preserves within it the summer half of the year permeated by the winter half of the year and vice versa. When yarrow is prepared in this way, the influences of an entire annual cycle become concentrated in it, and this gives rise to potentially new developmental possibilities.

During the course of preparing the yarrow, the stream of substances that end in seed formation in the flower are led beyond the restrictions normally imposed upon them by time and space. We intervene in the processes of nature and, as a result, what lives briefly in the flower as an open gesture, and in the seed as mere potential, becomes an active force on a higher level. The germinal-substance quality does not revert back to the natural processes of seed and humus formation, but instead germinates and is fertilised by the forces of the physical, etheric and astral bodies of the earth.

Effects of the preparation

The finished preparation, when measured against the quantities of ordinary manure used on a farm, is a seemingly negligible amount. What it contains, however, is not a finished work but a seed. Outwardly, it looks like humus and yet, by its nature and activity, it is precisely the opposite. Humus 'gives rise to a "lightless" working'.[8] The purpose of the yarrow preparation and, in a modified way, the other compost preparations too, is to bring a 'light effect' to the decaying organic materials arising on the farm. This enables these farm manures to become sensitive to the cosmic archetypes that exist beyond space and time.

The nature and meaning of the yarrow preparation is connected with the nature and meaning of potassium. In the living processes of the plant – and with yarrow this is archetypal – potassium increasingly liberates itself from the determinism of nature. It is able to rise a step closer towards its true cosmic nature because of the special sulphur configuration in the yarrow flowers. This may be called the level of 'active working'. In the field of tension that exists between sulphur and potassium in the living realm, potassium is active in the process that leads to the formation of leaf protein. This protein formation ceases almost entirely in the flower before continuing with the forming of seeds. In the flower, therefore, the potassium process is released from its living activity of protein formation. It refines itself in this germ-like state, expressed in the open gesture of the flower, and then begins to mineralise as the plant dies back, coming under the influence once more of the determinism of its physical properties.

It is this moment of flowering that the preparation seeks to preserve.

Through having been encased in the stag's bladder and undergone the further stages of preparation, the transformed potassium process that culminates in the flower continues to be active. The earthly substance of potassium is therefore not only raised into the living realm but into the sphere of sensation-bearing life. This vitalising and sensitising of substance also takes place in animals through their astral or soul nature. This is expressed by the remarkable fertilising power of ruminant manure, especially that of cattle. Cow manure is a good example of this.

The completed yarrow preparation represents the third stage of manuring. It becomes effective when homeopathically small amounts are added to compost and manure piles. The compost is then spread on the land at the appropriate time. The soil itself is only a very thin skin, forming a kind of 'diaphragm' between what is above. If the compost preparation is worked into it, this membrane-like 'middle zone' will be fertilised by this new substance. This manuring agent, which in essence consists of an inter-relationship between mineral, plant and animal elements, the rhythms of the solar year and the creative strength of the human spirit, enhances and exceeds the way the three kingdoms of nature work. It not only balances out what has been removed by the exploitation of the earth, but it gives the earth – the finished work of nature – the capacity to bring life to what is lifeless, enables what is living to bear sensations, and to become more individual through the spirit streaming into it from the future. Wherever it is carried out, the significance of this third step in the manuring process can be understood as being to lead nature beyond the threshold of space and time in order to reconnect it with the ongoing development of the human soul and spirit, and hence with the cosmos.

4: Chamomile (503)

Whereas the yarrow preparation works mostly with potassium, the chamomile preparation works largely with calcium compounds. It binds them together with other substances necessary for healthy plant growth and stabilises the nitrogen content in the compost. It also excludes the harmful effects of bacteria and fungi. Bacteria fructify the soil, causing decay and mineralisation, helping to transform the soil and build it up. But these processes are harmful when they occur above the earth, such as in the form of viral, bacterial or fungal plant diseases. We can see these diseases as soil processes working at the wrong level into the leaves and stems of plants, and we can counter them using compost treated with the chamomile preparation.

Chamomile's characteristics

The same steps are carried out for the chamomile preparation as for the yarrow, but in the text of the Agriculture Course the second step, that of hanging up the filled cow intestines used as a sheath, is not referred to. It is, however, mentioned in the notes on the course. We will explore this problem later on.

In its appearance, chamomile (*Matricaria recutita* or *Matricaria chamomilla*) is the polar opposite of yarrow, even though as members of the *Compositae* family they are closely related. However, a close connection is visible in that there is a functional polarity between the root and flower. It is this simultaneous closeness and distance that makes it the second in the series of compost preparations.

We know from the spiritual research of Rudolf Steiner that the chamomile preparation gives manure 'the power to receive so much life into itself that it can transmit life to the soil' and 'bind together, still more, the substances which are necessary for plant growth – that is, in addition to potassium, also the calcium compounds.' While yarrow offers an exceptional way of working with potassium, if we wish to work with the calcium influences we 'need another plant, which ... also contains sulphur in a homeopathic quantity and distribution, so as to attract through the sulphur the other substances which the plant needs.' It helps to 'exclude from the plant the harmful effects of fructification, thus keeping the plant in a healthy condition.'[1]

These indications throw light on the characteristic appearance of chamomile as compared to yarrow. The latter, thanks to the interaction between sulphur and potassium, expresses the full mastery of earth and water elements. Everything about the yarrow plant tends towards toughness of form and the restraining of life forces – as can be seen, for example, in the simultaneously firm and feathery leaves, and the bracts that serve to barricade in the tube-like flowers. By contrast, chamomile appears as though lifted away from the earth. Its relationship to potassium is revealed in its softly succulent leaves.[2] The activity of sulphur is focused primarily on the transforming and sublimating of calcium, and everything in its nature tends towards a strong subdivision of its vegetative organs. From the pronounced tap root that thickens, turnip-like, from below upwards, the side roots ray out horizontally and extend down and outwards. The leading shoot grows vertically upwards but very quickly produces numerous side shoots that radiate out and upwards, giving the plant a spherical form. The leaves have a loose, feathery, narrow, thread-like form, which only subdivide to a small degree (see Diagram 4.1).

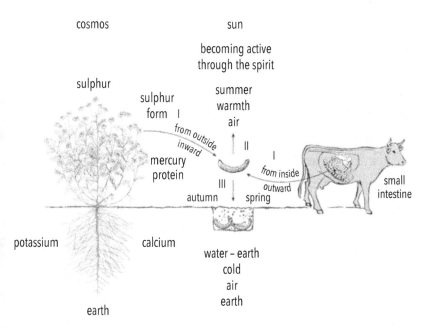

Diagram 4.1: Metamorphosis of substance during the stages of making the chamomile preparation.

These gestures of radiating and 'rounding off' point towards the rich variety of forms characteristic of limestone minerals ($CaCO_3$), such as the ray-like form of Aragonite or the rounded forms of crystalline chalk and stalagmites. The sprouting and subdividing form of chamomile reaches its terminus in the flowers. The stem growth is held back to form the flower receptacle upon which the tubular florets appear surrounded by white ray florets. As the flowers open, the receptacle grows into a ball and the tubular florets radiate a warm golden yellow. Beneath the convex form of the receptacle, an air-filled hollow develops – a characteristic that identifies the plant as true chamomile. A special indication of mobility in the chamomile, in contrast to that of yarrow, is the opening up of the ray florets in the morning and their closing at nightfall.

In the flowers, the light-, warmth- and air-permeated nature comes to expression on a higher level. The sulphur process is active throughout the plant in a different way to yarrow, including in the roots, and instead of keeping the potassium in a living state within the realm of earth and water, the chamomile etherises it along with the calcium as they stream upwards into the realm of warmth and light to culminate in the flower, where they release an intense aroma.

What comes to an end as a process in the flower is the starting point for something new in the chamomile preparation. This transformative turning inside out comes about when the harvested chamomile flowers are stuffed into an animal organ sheath, in this case a length of small intestine from a cow (see Diagram 4.1, II). The small intestine (*Intestium tenue*) starts from the duodenum, which exits the stomach, continues to the ileum and ends in the appendix. The long piece in between forms the jejunum, which is surrounded by the spiral form of the colon. This is the central part of the large intestine that extends between appendix and rectum. For the preparation it is the small intestine that is used. That is where the main digestion takes place, guided by secretions entering the duodenum from the liver, gall bladder and pancreas, and from the intestines' own glands. The surface area of the small intestine wall is greatly increased by tiny hair-like projections, or villi, that contain blood vessels and help absorb nutrients. Protected by a strong mucus layer, the blood and lymph vessels extend well into the intestinal cavity. This is where, due to gland secretions and bacteriological processes, the near complete mineralisation of ingested food material takes place. It is an intense metabolic process, accompanied by the rhythmic relaxing and tensing of the villi hairs, as well as the peristaltic movement of the intestinal wall through a layer of muscle that is separated from the abdominal cavity by a virtually transparent membrane (known as the serous membrane) thoroughly permeated by nerve tissue.

This division into three membranes and their three different roles also appears in the stag's bladder, and yet these two organs carry out opposite functions. The bladder collects and stores the liquids excreted by the kidneys and releases them to the world outside. The small intestine, by contrast, takes in the solid-liquid material of the food, breaks it down, sifts it as it passes through the intestinal wall, and then releases these sifted substances into the internal world of the organism. The bladder discharges from the body what the kidneys have sifted out as being unusable. The small intestine filters what is of value from food solids and releases it back into the organism.

Stages of the preparation

In the first stage of preparation we are once again faced with the task of carrying a particular process beyond the point at which it begins to fade in the plant so that it may remain active on a higher level. This task of liberation and transformation is fulfilled by the small intestine of the cow. Ideally we use one from a cow that has been reared on the farm's own home-produced feed. We cut up the small intestine into 25–50 centimetres (10–20 in) segments, tie one end together with string and with the help of a funnel, stuff the chamomile flower heads into the intestinal cavity. We then tie the end with string to produce thick sausages.

What were previously open to the cosmos and turned towards the sun as chamomile flowers, are now pressed into the inside of the small intestine of a cow. This change from the outside to the inside also applies to the new function of the intestine itself. It has a content that it no longer digests but instead preserves and makes receptive to the forces streaming in from the planetary surroundings through the sunlight. The normal function of the intestine is to transmit the processed food material it contains to the cow's internal organism via the lymph and blood streams. But now the outer skin of the intestine (the serous membrane of nervous tissue) relates directly to the forces of the cosmos and the earth. Freed from the cow's organism, the substance of the intestinal membrane transmits earth and sun forces to the chamomile flowers.

With the first preparation step of inverting polar opposite processes, an enhancement takes place that can be understood as giving the 'finished work' that appears in the plant the potential of entering a new stage of effective development. This potential is taken further in the second and third steps of the preparation (see Diagram 4.1, II and III).

With regard to the second step there is some uncertainty regarding the indications given by Rudolf Steiner in the Agriculture Course and those in the notes he used in preparation for it. In the American edition they are reproduced as an appendix. On page 30 of these notes we find the comment: 'Intestines – hang up'.[3] Nothing of this was mentioned in the fifth lecture when it was given. Far more emphasis was placed upon the third step, the exposure in the earth to the winter forces and the influences of nature: 'living vitality connected as nearly as possible with the *earthly* nature must be allowed to work upon the substance.'[4] A more exact description of the site was also given where the chamomile preparation should be buried:

> Therefore you should take these precious little sausages
> ... and expose them to the earth throughout the
> winter. Bury them not too deep, in a soil as rich as
> possible in humus. If possible, choose a spot where the
> snow will remain for a long time and where the sun will
> shine upon the snow, for you will thus contrive to let
> the cosmic-astral influences work down into the soil.[5]

It is this third stage that is of real significance for the chamomile preparation when compared to that of the yarrow. The second step, which concerns its exposure to cosmic influences via the air and warmth, is not referred to. In recent years, however, it has been commonplace to hang up the intestines over the summer months in the same way as the yarrow. The flower heads are gathered in May and June, partially dried, and stuffed into the intestines before the summer solstice. They are then exposed to the summer forces in the air and warmth.[6]

If in the spirit of the Agriculture Course, we try to evaluate the differing comments made regarding the effects of summer and winter forces in relation to chamomile, the following considerations can be helpful. The various ways of working with the preparation plants is determined by the characteristic way in which they work with cosmic forces and earthly substances. As has been described, the growth of chamomile is an expression of the intense astrality prevailing in the light, warmth and air. This increases still more in the flower and enters the air-filled hollow beneath the rounded receptacle (which may be understood as expressing the formation of an as yet unoccupied inner soul space), the diurnal movement of the ray florets, the warm, airy scent and its powerful healing property. All of this suggests that in its whole gesture, chamomile is subject to such a strong, summertime cosmic-astral influence that the second preparation stage seems a bit unnecessary – it has in effect already been completed. At the same time, however, hanging the chamomile sausages up over the summer will do them no harm.

Effects of the preparation

After overwintering in the earth, the chamomile is taken out of the ground in spring around Easter time. A new substance with new properties has now been created. According to Steiner, when this is added to the manure along with the other preparations:

> You will thus get a manure with a more stable nitrogen content, and with the added virtue of kindling the life in the earth, so that the earth itself will have a wonderfully stimulating effect on the plant-growth. Above all, you will create more healthy plants.[7]

The 'stimulating effect' of the chamomile preparation helps to counter the 'harmful effects of fructification' referred to earlier.

As with the other compost preparations, the chamomile preparation is best stored in earthenware containers surrounded by peat and kept in a dark place with an even temperature. Peat isolates and prevents loss by radiation. The preparation is applied in portions consisting of a pinch held between three fingers, or 2–3 grams per 2 cubic metres (70 cu ft) of manure or compost depending on the size of the heap. A right measure will be found by developing a personal relationship to the manure.

5: Stinging Nettle (504)

The stinging nettle preparation works with potassium, calcium and iron, sensitising the solid, earthbound nature of these minerals so that they bring healing and harmony to plant life. This preparation makes the soil 'inwardly sensitive', helping it to individualise its processes in relation to the plants being grown. According to Steiner, applying the stinging nettle preparation is like permeating the soil with 'reason and intelligence',[1] which is to say that the soil processes adjust and organise themselves as though they were being governed by a higher organism.

Stinging nettle's characteristics

The stinging nettle as the third member of the compost preparations group bears little resemblance to yarrow and chamomile in terms of its appearance. But it is similar with regard to the way it works with potassium, calcium and also iron. The sulphur process permeates the entire plant from above, giving it the capacity to draw the substances referred to out of its earthly-inorganic nature and bring their living processes to manifestation.

Rudolf Steiner describes the nettle, which belongs to the *Urticaceae* family, as 'the greatest benefactor of plant growth in general', and he goes on to say:

> The stinging nettle is a regular 'Jack-of-all-trades'.
> It can do very, very much … Moreover the stinging
> nettle carries potassium and calcium in its currents
> and radiations, and in addition has a kind of *iron*
> radiation. These iron radiations of the nettle are almost
> as beneficial to the whole course of Nature as our own
> iron radiations in our blood. Truly, the stinging nettle is
> such a good fellow and does not deserve the contempt
> with which we often look down on it where it grows
> wild in Nature. It should really grow around the human
> heart, for in the world outside – in its marvellous inner
> working and inner organisation – it is wonderfully
> similar in the way it works to the way the heart works
> in the human organism.[2]

The strict order of its upright form and the intense green of its leaves, though an effect of nitrogen, demonstrates that the nettle is a master of the iron process, as without iron the magnesium-rich chlorophyll could not form. This is in turn an expression of an enhanced human 'I' force. Thanks to its 'marvellous inner working', the nettle brings order wherever it grows. It brings one-sided soil processes into harmonious balance while forming extremely stable and crumb-like humus from the waste.

What is striking is that the strong polarity we find in yarrow and chamomile between the root and flower, is absent in the nettle. Its insignificant flowers are found among the leaves in the upper third of the stem. The flowers develop from the axles of the leaf pairs, which form at right angles to one another (see Diagram 5.1). The leaves of the pollen-bearing plants have a more elongated form and more rounded serrations along the perimeter. The seed-bearing plants on the other hand have more compact, heart-shaped leaves and sharply pointed serrations. Metamorphosis of the leaves is very limited. The lowest leaf already reveals the plant type with well-developed if somewhat more rounded serrations and a contingent of 'stinging' hairs – even the embryonic leaves have them. Up towards the middle of the stem the leaves widen and develop a pronounced heart shape, and further towards the top of the stem they get narrower and once again take on a fine and pointed lance-like form.

The crowning moment of flowering plants – the blossoming of the flower itself – is almost non-existent. The flowers have no petals. The yellow-green pollen flowers grow into long trusses and hang down within the upper portion of the plant. The seed-bearing flowers are whitish-green and are pressed together close to the stem.

The leaves, the four cornered hollow stems, as well as the flowers, are covered with stinging hairs. These are single-celled outgrowths whose bases are strengthened with a calcium compound. Their tips have heads of silica that break off when touched, releasing the poisonous cell juice urtizine, which is what does the stinging. This contains amongst other things substances that are normally found only in animals and humans, such as histamine, serotonin and acetylcholine.

Through the stinging hairs the plant dies towards the periphery in a physiological process that is characteristic of the flower. We can therefore say with some justification that not only are the flowers of the stinging nettle drawn down into the leaves, but the entire plant, as it grows above the earth, is clothed in a kind of flowering process from its earliest moments onwards.

The stinging nettle grows from a mass of roots, also known as rhizomes, to form colonies 1–2 metres (3–6 ft) in height. The runners are yellowish in colour and run just beneath the surface of the soil. From their nodal points they send down tough yellow root strands, and from these develop fine, white fibrous roots that branch out in the upper soil. Two plantlets grow vertically from each node on the rhizome. They strive upwards and have a strict harmonious geometry. Growing island-like, the shoots form a protective shield around an inner space heavily shaded by the leaves.

The wonderful thing about the stinging nettle is that even the polarity between root and flower is enhanced to a higher unity. The sulphur process working from above and the salt process working from below remain highly mobile, permeating and enhancing one another in the leaves. On a higher level of development this process can be compared to the circulation of the blood. In the heart, midway between what is above and below, the venous and arterial blood streams meet. Arterial blood is carried out to the periphery, returning as venous blood to the heart where, via the breathing of the lungs, it is transformed once more into arterial blood. This also occurs on the level of the plant in the stinging nettle – it 'flowers' from out of its centre to the periphery of the whole plant.

The earthbound substances, potassium, calcium and iron, do not just experience their 'purification', their etherisation, in the culmination of the flowering process and the dying into form. They also experience the flowering process of the nettle, the entirety of its growing activity influenced by the astral forces that work through sulphur, in such a way that mobility is retained. As a result the stinging nettle is able, through 'its marvellous inner working', to be 'similar to what the heart is in the human organism.' This is most likely the reason why Rudolf Steiner did not suggest an animal organ for use in making the stinging nettle preparation, for the organ in question would be the heart. The heart, however, is not a membrane but a muscle.

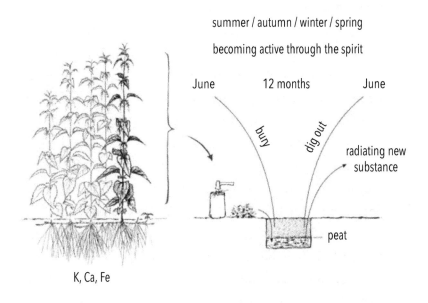

Diagram 5.1: Preparing the stinging nettle preparation during the course of a year.

Stages of the preparation

In the first stage of the preparation, the nettles are cut, bent together or roughly chopped, and allowed to wilt a little. A shallow hole is dug in the topsoil, the material is put in it surrounded by some peat, then covered with soil. It remains there for a whole year, exposed to both the winter and summer forces. After one year, the material is dug out, which, thanks to the advanced humification process, is now greatly reduced in quantity. In our hands we then hold the finished preparation.

Some skill is required to achieve just the right amount of wilting. Direct sunlight should also be avoided. If the stinging nettle material is too moist it can easily start fermenting in the wrong way, and if the light is too strong it turns black. It is best gently dried in the shade and best of all with warm air passing through it. In order to find it again a year later, use is often made of old hessian sacks, flat and thin-walled wooden vegetable boxes, or drainage pipes into which they are stuffed. In the latter case we end up with largely humified black material.

What applies to the other preparations buried in the earth is particularly important for the stinging nettle. Whether the preparation is more or less successful depends not on the degree to which it has become humified, but on how loose the structure of the preparation is, how pleasant its smell is, and above all whether the surrounding soil is in a good, well-aerated, crumbly and living condition. All of the compost preparations with the exception of valerian are placed separately in the compost or manure pile in 30–50 centimetre (12–20 in) deep holes spaced 50 centimetres (20 in) to 3 metres (10 ft) apart, depending on the size or length of the heap. The stinging nettle preparation with its heart-related function is generally inserted at the centre of the pile; the four sister preparations are then placed appropriately along the edges or corners. Their effects are of an astral radiant nature, one whose quality can be perceived, but not quantitatively measured.

Effects of the preparation

The biodynamic preparations are substances that 'fertilise', each in its own way, the relationship between the cosmos and the earth. This is where the yarrow, chamomile and stinging nettle preparations take on particular significance. They bring about a special kind of transubstantiation whereby the earthbound substances of potassium, calcium and iron gain properties that approximate those of nitrogen. Rudolf Steiner gives the following indication:

> For there is a hidden alchemy in the organic process. This hidden alchemy really transmutes the potassium, for example, into nitrogen, provided only that the potassium is working properly in the organic process.

Furthermore there

> ... is a mutual and qualitative relationship between the calcium and the hydrogen, similar to that between oxygen and nitrogen in the air ... under the influence

of hydrogen, calcium and potassium are continually being transmuted into something very like nitrogen, and at length into actual nitrogen. And the nitrogen which is formed in this way is of the greatest benefit to plant-growth. We must enable it to be thus engendered by methods such as I have here described.[3]

This means working artistically with processes occurring in the farm organism that create substances. Just as the painter uses canvas, colour and brush as tools for their artistic creativity, so does the farmer use and vitalise the earth by making and applying the preparations.

How can we begin to solve this riddle of transformation or transubstantiation?

Certainly not by applying the same way of thinking that we use to unlock the mysteries of inorganic nature. The two qualitative relationships quoted above are opposite to one another. Oxygen and nitrogen are major constituents of the air we breathe; both are found in a gaseous, inorganic and inactive state and surround the growing plant from the outside. Potassium and calcium behave differently. Through the living processes in plants their spell of enchantment as physical substances is broken, and they are lifted up into the sphere of living activity. This can be understood in the following way. Medicinal plants like yarrow, chamomile and stinging nettle have the capacity, through the strength of their etheric body, to release the physically dense substances of potassium and calcium from their imprisonment in form and bring them into movement. In their appearance and in their properties they reflect an earlier stage of evolution when they existed in a living condition. Since the mineral ascends into a process of life in these plants, we may construe that potassium and calcium become receptive in a new way to the activity of their next higher member, namely the etheric body of these substances. The reality of this level of being is, according to spiritual research, to be found in the lowest realm of the spirit that borders on the physical-sense world. New evolutionary soul impulses from the future can then enable these substances, now with their own etheric body, to open themselves up to qualities of being that are at home in still higher spiritual realms. The substances of potassium and calcium by this route become 'like nitrogen' in that they become bearers of soul forces. They acquire the properties of nitrogen, which bears the astral world of archetypes into the processes of life. Through the formation of protein, nitrogen brings processes of life into the living world that originate in the earth's distant past, have gone through various stages of development, and have now become the 'finished work' of creation.

Hydrogen is also present in the soil. The special qualitative relationship potassium and calcium have to this element allows them to be led back to their etheric condition. While these salt-forming substances have entered most deeply into purely terrestrial conditions, it is hydrogen that is 'nearly as possible akin to the physical and yet again as nearly akin to the spiritual.'[4] It makes use of sulphur for building up the organism as well as for breaking it down, it 'carries out again into the far spaces of the Universe all that is *formed*, and *alive*, and *astral* ... It is hydrogen which dissolves everything.'[5]

Now it is precisely this qualitative relationship referred to in yarrow, chamomile and stinging nettle that enables the hydrogen to dissolve the structure that binds potassium and calcium and release the astrality congealed within it into the 'undefined chaos of the universal All'.[6] Through these plants the etheric nature of potassium and calcium can be brought into movement again so that through the mediation of sulphur, it can become the bearer of new forces. Potassium and calcium are transformed into nitrogen to the extent that they become bearers of a new astrality. It could be said that the nitrogen of the air replicates the past in the present, whereas the transformed potassium and calcium, as bearers of astral forces, enable the future to be brought into the present.

How can this same substance – nitrogen – have such different qualitative effects ascribed to it? Current scientific understanding considers only its inorganic, physical-chemical properties. If this approach is deemed to be the only valid one, it is justified to ask why we should go to such complicated lengths to produce such tiny amounts of nitrogen when there is a surplus of it in the air. After all, in order to provide the nitrogen required for consistent growth, all that is required is to integrate sufficient legumes in the arable crop rotation and have enough organic manure to hand. This view only explains a portion of the reality. If the physically entranced forces are brought to life in the plant through its etheric organisation or ensouled through the animal's astral body, they become mobile; they are no longer subject to purely physical laws but those of the living etheric and formative astral worlds.

The transformation of dead, inorganic matter brought about

in this way enables plants, animals and human beings to incarnate on the earth. These incarnations repeat developments of a far distant past; they are processes determined by iron necessity. It is only human beings who, having incarnated souls-spirits, are able to determine their own future development in freedom. The upbuilding creative forces of the body are thus continually being spiritualised or transubstantiated by the 'I'. This indicates the true future direction of human evolution.

In the Agricultural Course, Rudolf Steiner describes how, by engaging freely in the making of the biodynamic preparations, manuring substances are created that render the 'finished work' of nature receptive to evolutionary impulses that bring forces of the future to manifestation in the present.

In the case of the yarrow, chamomile and stinging nettle preparations, potassium, calcium and iron are subjected, with the help of hydrogen, to a transformative process through which they become like nitrogen, the carrier of astral forces. It is a process of transubstantiation through which the boundaries set by nature are transcended. Seen in this light, the way this form of nitrogen works must be understood as being the polar opposite to nitrogen as it normally appears in nature. The nitrogen produced through the preparation of yarrow, chamomile and stinging nettle is one that is of 'the greatest benefit to plant growth.'[7] As a form of manuring, the first three preparations work with the developing human soul to bring about a new potential for the development and transformation of nature. It can only serve the good if it is also accompanied by a great soul transformation in the human being. Where this doesn't occur it will cause destruction.

6: Oak Bark (505)

The oak bark preparation helps to combat plant diseases by working with calcium to restore order to the etheric bodies of plants. When the etheric body works too strongly it can cause rampant growth, which is what brings about disease. Calcium helps to dampen down this activity and improves the ability of plants to hold themselves together. According to Steiner, the structure in which calcium exists in oak bark is particularly well suited to this.

Oak's characteristics

In temperate latitudes the English Oak (*Quercus robur*) represents the world of perennial woody plants. In order to consider the unique qualities of the tree in a pictorial way, we must first look at the slow and gradual coming into being of its mighty, gnarled form, and to the substances that manifest on the one hand in the bark and on the other in the hard and resistant heartwood.

This process begins in the living layer of cambium and becomes more mineralised, both towards the centre of the tree and on the outside, as it nears the end point of living growth. Spiritual research indicates the importance of the calcium process in the oak bark and how it keeps the plant healthy. It 'restores order when the ether-body is working too strongly.'[1] In the living tissue beneath the bark, an organic calcium compound called calcium oxalate forms in certain cells alongside the tannins and other aromatic substances. This compound crystallises in the cell vacuoles and forms rounded crystal clusters in the bark, which, being hard to dissolve in water, is long lasting.

Once it has germinated, the oak sends its tap root deep into the ground followed by side or heart roots that go down equally deep. From these grow powerful, wide-spreading lateral roots, which in turn send so-called sinker roots into the depths around the periphery.[2] When fully grown the oak forms a 'crown' of roots in the earth that reflects the crown of the tree above. Its entire growth pattern, and with it the chemistry of its substances, point towards its connection to the earth. The young oak shoot grows up vertically with many side shoots, and in its growth habit it is similar to other broad-leaf trees. But the unmistakeable archetype of the oak is revealed early on in its life by the very characteristically indented and lobed leaves. It is only after about twenty years that the appearance of the oak changes and takes on the familiar form of a broad, wide-set, irregular, open and light-filled crown. What was indicated early on by the leaves now manifests itself in the power of the whole tree as it grows during the succeeding decades. The powerful growth force that persists even in ancient oak trees gives the impression of every branch being dammed up or concentrated, showing secondary growth rather than new shoots. This tendency shows itself in the hard and long-lasting wood of the trunk and boughs. The leaves, too, are concentrated in clusters. With increasing girth, the earlier silver bark – so-called because of its smooth shininess – disappears after about twenty years and is replaced by the characteristic, deeply fissured bark.

Rudolf Steiner referred to the particularly ideal structure of calcium as it is found in oak bark, although the different botanical concepts of bark were not mentioned.[3] This led to differing approaches and debates over whether silver bark or mature bark should be used for the preparation. In this book we are referring to mature bark because silver bark disappears as soon as the oak attains its typically mature form. The transition to mature bark involves the formation of cork cambium whose cells protect the living layer from external influences and convert the dying tissues into the cork-like substance of mature bark. The mature bark is thus made from layers of annually extruded, cork-like tissue that remain tightly held together.

As the trunk and spreading branches thicken, the bark develops deep fissures. The cork cells prevent the decay of the dead tissue, as well as the calcium oxalate, some of the heavier volatile oils, and the aromatic compounds that are formed in the bark along with their derivatives. The aromatic compounds are hydrocarbon substances whose end product is the highly polymerised resin. These compounds include, amongst other things, volatile aromas akin to the scents of flowers or the aromatic hydrocarbons contained in plant oils. These arise during the devitalisation phase as a plant comes to flower, when they either 'vaporise' as scent, or are retained for a longer period. The latter is the case with bark, which is less volatile and surrounded by cork-like material. The 'flower-like' composition of these substances in the bark is due to a decline in the vitality of its formative forces. Life is driven out of the bark just as it is during the process of flowering. Only now does the bark of the oak have the structure to enable the calcium to develop its plant-healing properties and, during the stages of preparation, make it useful for both soil and plant. The oak only begins to flower when its tree crown is fully developed. This occurs when it is about sixty years old and the flowers are easily overlooked. It is only in September and October, when the acorns fall from a tree of this age, that we become aware of the fundamental change that has occurred. Linking this to our earlier observations, we might wonder if the oak has been flowering and bearing fruit long before the first visible flowers and fruit appeared? Trees of all kinds are described by Steiner as being like mounds of raised-up earth, and the oak is especially representative of this.[4] In this raised-up quality we can perhaps see a kind of flowering and fruiting process taking place at a half-way stage of development that is, however, held back in the trunk and branches.

If we look at the trunk of an oak in cross section, we find expressed in its horizontal rings the three alchemical principles of Paracelsus, known as Sulphur, Mercury and Salt. These do not refer to the substances of the same name, but rather to processes. The Sulphur process refers to an expansive force, to evaporation and dissolution; the Salt process refers to the power of contraction and crystallisation, and the Mercury process is that which mediates between the two, being both fluid and possessing form. In alchemy, these processes can involve different substances. Returning to the cross section of the oak trunk, we find three layers: the outermost bark, a thin layer of cambium and then the xylem.

The xylem, the actual wood of the tree, with its network of tissues that carries water and salts, 'takes root' towards the centre in the sapwood. This is where assimilated substances are stored and mobilised again, as occurs in the resorption of nutrients in the soil. The sapwood dies towards the centre of the trunk; it becomes mineralised as heartwood through the depositing of woody substances that prevent decay. This is the zone of the Salt process.

The outside layer, loosely called the bark, contains the phloem network. This carries everything that has been assimilated into all the green parts of the tree. It is bounded by the fibres of the cork-like bark, which gives protection against the wind and the weather and gives form to the plant's appearance. The bark conceals numerous cell tissues, some of which contain chlorophyll that absorb light, as well as single cells that form calcium oxalate crystals. There is an intense yet restrained form of metabolic activity at work here throughout the bark and flower-like substances, forming aromatic compounds such as tannins and their derivatives. This layer corresponds with the Sulphur process.

The living layer of cambium is hidden beneath the bark and surrounds the tree like a green leaf. In this thin living layer beneath the bark, we can identify a held-back Mercury process that does not differentiate itself into leaves and shoots. The leaf-sprouting nature of the living bark is evident from the leaf buds that emerge from this layer, and the growing shoots that are in continuous connection with the cambium. The mineral ingredients, and the coming into being of hydrocarbons through a form of sulphurisation, points towards the existence of this held-back flowering process that is still directed by the forces of growth. The living bark is connected with the heartwood by piths that ray out through the cambium.

Just as the stem and leaf in a plant metamorphose into the flower, so does the living bark transform into mature bark. This metamorphosis occurs in such a way that a secondary layer of cambium, known as cork cambium, is formed. The resulting cork tissue connects and holds together the continually dying outer layer of bark as well as the mineral ingredients and the less volatile aromatic compounds. Just as the fully developed flower dies into form and colour and into volatile aromatic substances, which give it its scent, so too does the leafy growth of the oak become bark. In this way everything that was previously permeated with life becomes stark and dead in form, what was green becomes reddish, earthy brown and what were highly volatile aromatic compounds dissipate.

From this point of view, we can say that a kind of flowering process does in fact occur in the mature bark, but one that remains incomplete and on a lower level. It does not result in a complete transformation of earthly substance as it does in the flower, but in a preservation of the organic-mineral substances that originated in the living bark.

In observing a bark-producing tree like the oak we can say that the mature bark represents a kind of perennial flowering process that takes place close to the earth. Along with the living bark, cambium, xylem (sapwood and heartwood), this bequeaths a certain permanence to the tree as the seasons change. This means that the starting material for the oak bark preparation is also a substance that comes from a flower-like process. This underscores the answer that Rudolf Steiner gave to the question whether the entire bark should be used: 'No, only the surface – the outermost layer of bark which crumbles off of its own accord when you loosen it.'[5]

The oak possesses a strong power of attraction for insects. Among the roots and debris of hollow trunks we find the larvae of numerous beetles, and among the leaves we find gall wasps, whose home is the gall or oak apple. The oak is host to more than a hundred different species of gall wasps. The crown of the tree has a strong relationship to sentient animals. Only by observing the oak comprehensively – at the specific way it processes substances, its relationship to the insect world and the forces of its surroundings – can we understand what Rudolf Steiner meant when he said, 'the calcium structure in the rind of the oak is absolutely ideal',[6] when prepared correctly, to produce a manure that can prevent disease prophylactically.

The skull of a domestic animal

If the oak stands out as unique among the preparation plants, so too does the animal organ sheath that is used for preparing the oak bark – the skull of a domestic animal.

According to Steiner, 'the skull of any of our domestic animals will do – it makes little or no difference.'[7] This statement conceals a great riddle. In the case of the other preparations the organ sheath is taken from a very specific species of animal – red deer stag or domestic cattle. However, in the case of the oak bark preparation, the kind of animal is not important, only that it fulfils the role of a domestic animal. These vary greatly and distinguish themselves significantly from their wild relatives in terms of form and behaviour. What is it then, transcending of species, family and order, that makes an animal into a domestic animal?

Domestic animals exist thanks to the intervention of human beings. This is not simply a question of breeding in the way it is understood today, but of education: the domestic animal needs an education to become a domestic animal just as the human being needs one to become human. This educating of the animal into domesticity in a conscious and species-appropriate way is an art that enables the animal to grow beyond its inborn instincts. The domestic animal renounces, to some extent, the wisdom-filled instincts of its wild form. Human beings then have the responsibility to make up for this loss. Because animals have no personal self, they need the guidance of human beings. With the disappearance of traditional farming practices and the connections between animals and humans that formed part of them, a new understanding of animals is needed today that gives value to the working animal. Biodynamic animal husbandry is founded on this approach. If a deepened understanding of the nature and development of domestic animals is sought in this way, then it will be possible to find skulls that are suitable for making the oak bark preparation. At present, it is common practice to use the skull of a cow. If slaughtered in the autumn, the cow can also provide the organs for the chamomile and dandelion preparations. Sheep and goat skulls can also be used if necessary.

There are two parts of skull: facial skeleton (or viscero-cranium) and neurocranium or braincase. New-born animals have a unified, almost round form similar to that which humans have throughout their life. The facial skull then grows longer and becomes dominant during the animal's brief juvenile phase. At this point the head risks being overpowered by metabolic forces. This phenomenon, and how it is overcome, can be seen most clearly in the antler-bearing animals and, in a different way, in cattle. The stag, for example, achieves a major metabolic feat each year in that antlers permeated with blood and coated with a velvet skin grow out of its brain skull. This occurs during the first half of the year. At the beginning of the second half of the year the metabolic power shooting up from the head is stilled and the bony structure dies back, leaving an inner space surrounded by the branching antler antennae. These become a mighty sense organ for sensing the warmth and light of the stag's surroundings. In the winter they are discarded. The same process occurs in an opposite way with regard to the horns of a cow. These grow continuously year on year while simultaneously dying into the horny exterior. Through this inwardly directed sense organ, the powerful metabolic processes pushing out towards the nerve-sense organisation are driven back into the body by the dead casing of the horn. The cow can therefore preserve, in a different way to the stag, the nerve-sense forces in the head from being encroached upon by the driving power of metabolism.

The anatomical development of the body and skull of domestic animals is no different to those of their wild equivalents, yet there is a significant difference in the way they are formed. The facial skeleton stays somewhat shorter and the size of the braincase is significantly reduced, sensory capacity is less and metabolic activity is greater. These are all symptoms of a development held back in the domestic animal, of youthful forces retained in a more embryonic state. This preserved youthfulness is what distinguishes the domestic animal; human beings are responsible on an evolutionary level for this youthfulness. This fact makes us duty bound to train, manage, feed and care for our domesticated animals out of knowledge and with love.

No conclusive answers can be gleaned from the foregoing considerations as to why Rudolf Steiner recommended a skull from any domestic animal for the oak bark preparation. A likely answer is to be found only by focusing on the relationship of humans to animals since the last ice age. Humanity at that time lived in a dream-like consciousness from which flowed the folk myths and inspirations that came from a supersensible spiritual world still experienced as real, and they were led by the inspired priests of the mystery centres. This is the spiritual background against which the origin of domestic animals must be sought. It involved gradually replacing what had evolved in the animals as instinctive life with the guiding leadership of human beings. This transformative step, arising from the soul-spiritual consciousness of people at that time, was impressed upon the life bodies of the animals and from there to their physical bodies and the line of inheritance.

This helped domestic animals to retain their youthfulness and keep their bodily constitution open to variation.

We may ultimately assume that the mystery of the physical coming-into-being of domestic animals, the specific arrangement of substances and their structure, came to expression where life had completely congealed into form. Once form is created, life withdraws or dies. The arrangement of substances created by the etheric forces of the domestic animal is different in the long bones of the limbs, in the bones of the pelvis or the backbone, and in the skull bones that enclose the centre of the nervous system. The main substance building the latter is calcium, which appears in various compounds with phosphorous, carbon, oxygen and fluorine. Its forces, focused towards a centre point, give the skull its almost round form. The compositional arrangement of calcium in the bones of the skull is an expression of the soul forces in the domestic animal that have been kept young by human engagement. The arrangement of calcium substances in animals is of a higher nature than that which occurs in plants, for example in the bark of an oak tree. Seen in this way, the skull of a domestic animal preserves the astral forces in its calcium compounds that the animal has experienced flowing towards it through the daily care it has received in the place of its lost instincts. In a higher sense, they have the capacity to work in a purifying, corrective and healing way on rampant life processes.

This attempt at understanding oak bark preparation and the skull of the domestic animal needs to be developed further. This does not however yet solve the riddle of this fourth preparation. It must be put into practice by human hands – an artistic act!

Stages of the preparation

If possible, the oak bark is obtained in September from one of the mature oaks growing on the farm and a skull is taken from one of the animals also kept on the farm. In the case of old oaks with deeply furrowed bark, it is worth scraping off the outermost and frequently moss-covered layer with a surform tool and only using the younger layers of bark beneath it. The partly crumbly mass is then completely broken down so that it has a good crumb structure.

After the animal has been slaughtered, the brain is removed from the skull via the rear orifice along with any remnants of flesh and skin. The skull cavity is then filled with ground oak bark through the same opening in the occipital bone through which the nerve fibres of the spinal cord entered the cerebellum. The opening is then closed with a splinter of bone and sealed with clay.

Through this initial stage of preparation the first inversion occurs. Something external, the bark, becomes something internal, enclosed by a bony receptacle, which even in the living animal was closer to death than life.

In the second stage there is another inversion. After being filled, the skulls are immediately buried in the earth where there is a lot of vegetable matter and where rainwater and snow can flow over it. In this earthy, watery environment they remain exposed to the autumn and winter forces.

Many biodynamic practitioners have found there are practical difficulties with meeting these conditions in an optimal way.[8] But a flexible approach suggested by Rudolf Steiner makes it easier.

> We lower it into the earth, but not too deep. We cover it over with peat-moss, and then introduce some kind of channel or waterpipe so as to let as much rainwater as possible flow into the place. (We might even do it as follows: take a barrel where rainwater is constantly flowing in and out. Put in it vegetable matter such as will bring about the continued presence of some vegetable slime. Let the bony vessel which contains the crumbled oak-bark lie in the slime in the water.)[9]

Either method can be practised.

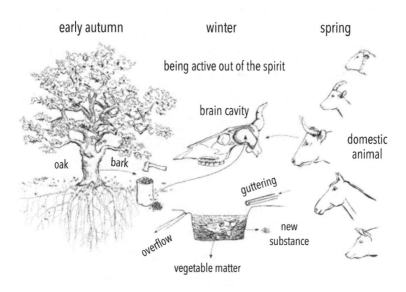

Diagram 6.1: The process of making oak bark preparation during the winter half of the year.

In spring, the skulls are removed from their earthy-watery domain. The oak bark has taken up and concentrated the forces working on it from the outside, and what appears is a new substance with a somewhat darker colouration and a slightly earthy yet still crumbly consistency. It is thus a new substance gifted with a new capacity that can give compost and manure the power 'prophylactically to combat or arrest any harmful plant diseases.'[10]

Effects of the preparation

How can we best think of this process, which plays itself out in the earth during the winter between the vegetable slime, the bony skull and the oak bark?

First we must consider how calcium's capacity to draw the etheric together can be put to use. All too often we find that a sudden spell of warm, sunny weather following a damp winter and spring causes masses of harmful organisms to appear, for example colonies of aphids. They are a sign of rampant etheric growth forces. The cosmic astral forces are too weak to keep these excessive growth forces in check. This is where the contracting forces of calcium can help, but not the acid remnant of carbonic acid $(CaCO_3)$ that is normally used. It needs a form of calcium that has structure, such as appears in the bark of the oak. This calcium oxalate compound, which has been lifted up by the life processes of the oak and then excreted into the bark, must first be brought through the preparation-making process to the point where, as an added ingredient to organic manure, it can keep the living relationship of plant, soil and the in-streaming astral forces of the inner planets, including the moon, in a healthy balance. It can gain this capacity through the creation of an environment of rampant etheric forces, a condition found in vegetable slime, that must be continually kept in check through the regular inflow and outflow of rainwater and snowmelt. The atmospheric water contains oxygen, which ensures that the anaerobic decomposition process does not flip over into putrefaction.

In contrast to the fluid nature of the vegetable slime, the inflowing rainwater and snowmelt is permeated by cosmic forces. These condense from the gaseous state of the air in the wintry atmosphere, forming droplets or crystallising as snow. The forces of the macrocosm then connect with these and concentrate themselves in the watery envelope. If these water droplets and snow crystals then coalesce as homogenous water masses in the earth, the cosmic forces concentrated in them go into the earth realm and, in this case, into the primal chaos of the vegetable slime. A kind of synthesis occurs that brings about balance. The moon-like microbial decomposition processes are brought into balance by the activity of current cosmic forces. We can assume that it is the coming together of inner and outer planetary influences that brings about this equilibrium.

Into this environment of formless, chaotic vegetable slime and the instreaming of rainwater and snowmelt from above, is placed the domestic animal skull whose brain cavity is filled with oak bark. A combination of etheric forces continuously stream around the skull and penetrate the bony, cranial vault. In it is a calcium structure formed by the forces of the domestic animal's higher astral nature. From the heart, which is related to the sun, towards the head of the animal the forces of the outer planets Mars and Jupiter are working, while in the head itself the form- and structure-building forces of Saturn dominate.

This means that the etheric forces in the vegetable slime, which were brought into balance by rainwater, are turned into formative forces as a result of their passing through the skull casing by the less instinct-bound astral forces of the domestic animal. The calcium of the skull roof that was given structure by the astral body of the domestic animal, draws the ether forces together and gives them the power to build up and bring about order. In the oak bark, the calcium is prepared in such a form that on the one hand it becomes sensitive to the etheric stream of the vegetable slime, which has now been transformed by the domestic animal skull into formative forces, while on the other hand it concentrates and preserves these newly received formative forces in connection with the activity of the oak bark.

The domestic animal skull and its contents remain exposed throughout the winter to the terrestrial-cosmic forces streaming into this earthy-watery realm. This can be understood in that the activity of the inner planets, and especially the moon, is strongest in the watery context of vegetable slime during this time, while the forces of form and structure, which emanate from the distant realm of the fixed stars that work through the solid earth element, unfold their greatest activity in the hard and dense skull of the domestic animal.

The finished preparation is then inserted in the compost or manure pile at the rate of one pinch per 1–2 cubic metres (35–70 cu ft) in the garden or 8–15 cubic metres (300–500 cu ft) on a farm. It is not the material that is important, but the radiation of forces. Again, it is about learning through personal observation and experience.

As with all the preparations, each of the steps are not additive in effect but multiplicative. Substances arise whose properties are not variations of pre-existing qualities from the past but are properties that enable developments to take place in a living context. In the case of the oak bark preparation, the new property is the ability to strengthen the parts of the plant and thereby counteract or prevent plant diseases.

The natural substances and forces that are brought together in a new way to produce such characteristics are the following:

1. The oak bark. The flowering process of the oak is more earthly and this expresses itself in the bark as calcium oxalate.
2. The skull of a domestic animal. This metamorphoses under the guidance of the human being. The wild animal is formed by the spirit of the group soul, in the domestic animal that role is performed by the spirit of the human being.
3. The skull cavity, in which the brain is held surrounded by brain fluid, is a space lifted out of direct contact with the earth and forms a microcosmic image of the cosmos.
4. In the winter the earth is spiritually awake.

In these four qualities a centripetal, earth-focused tendency expresses itself. They appear naturally separated. If they were brought into connection with each other, both spatially and temporally, on the foundation of vegetable slime and the in-streaming atmospheric waters, calcium, with its new qualities, would be able therapeutically to balance out one-sided terrestrial qualities.

7: Dandelion (506)

The dandelion preparation works with silicic acid, but not the kind that is derived from quartz and silicates. According to Steiner, the relevant form of silicic acid is one that is finely and homeopathically distributed in the surrounding cosmos. The dandelion helps to draw this in and relate it in the right way to potassium.

Dandelion's characteristics

In looking for a plant that can bring about the correct relationship between silicic acid and potassium, Rudolf Steiner recommended the dandelion (*Taraxacum officinale*).

> The innocent yellow dandelion! In whatever
> district it grows, [the dandelion] is the greatest
> boon; for it mediates between the silicic acid finely,
> homeopathically distributed in the cosmos, and that
> which is needed as silicic acid throughout the given
> district of the earth. Truly this dandelion is a kind of
> messenger of heaven.[1]

The dandelion has evolved so that in the right way it can 'draw in silicic acid from the entire cosmos':

> But we shall also need something to attract the silicic acid from the whole cosmic environment, for we must have this silicic acid in the plant. Precisely with regard to silicic acid, the earth gradually loses its power. It loses it very slowly, therefore we do not notice it.[2]

Steiner goes onto say that this silicic acid is of the greatest importance for plant growth.

> Silicic acid contains silicon ... and silicon, too, is transmuted in the living organism – transmuted into a substance of great importance, which, however, is not yet included among the chemical elements at all.[3]

Once again, Rudolf Steiner points towards a riddle, the solution to which can only be found by observing certain physical phenomena that suggest a higher context. That context can only be accessed with a thought-through experience of the spirit.

The dandelion possesses a wealth of characteristics that point towards a uniqueness that, even within the *Compositae* family, is unparalleled. It flowers in springtime, when the earth emerges from her winter rest and breathes out once more into her surroundings. The profusion of dandelion flowers that appear in meadows and pastures and along roadsides in early April is an image of this reconnection of the earth and the cosmos. Then, as quickly as it appears, it vanishes from sight amidst the burgeoning growth of the accompanying grasses and herbs. A few of the golden flowers will appear again in early autumn. For the rest of the year the dandelions concentrate and store up forces from the cosmos and the earth, which then, almost overnight, grow as wave upon wave of flower stems in the ascending radiance of the spring sun. The flower buds are borne upwards and then open out to reveal a prominent receptacle and a head full of radiant florets that turn towards the sun.

As a perennial herbaceous plant (it lives for about eight years), the dandelion is thoroughly dominated by the formative astral activity at work within it, in a way that is comparable though opposite to the stinging nettle. While the stinging nettle presents itself assertively to the world and protects its own grandiose inner activity, the dandelion demonstrates in its whole being how it is totally given over to both the earth and the cosmos. Although the astral organisation of the dandelion holds the ether forces powerfully in its grasp, its gesture is one of selflessness and of desiring nothing for itself – 'the innocent yellow dandelion' as Steiner called it.

The root

Already in the root, the polarity between contraction and the living power of expansion is powerfully expressed. This shows itself in the strength with which the tap root plunges into the earth and branches out in the humus-rich topsoil. The roots penetrate deep into the mineralised sub-soil where the fine root hairs spread out. Even segments of the upper part of the tap root can grow to mature plants.[4] The upper sections of the branching primary root thicken in the topsoil to become a turnip-like root rhizome. This is made up of a loose conglomeration of cells permeated with a network of interconnected tubes. These channels carry a viscous, milky-white juice, which is under pressure. This shows the effect of contraction right up into the flower stem. If the root, leaf rib or flower stem is cut, the milky juice immediately oozes out. The entire dandelion plant is continually under pressure. One could also say that the etheric body of this plant is subject to permanent 'formative pressure' from the astral.

Shoot and leaf

Compared with the 'Salt' pole of the taproot and transitional root rhizome, the 'Mercury' middle realm of the dandelion's shoot and leaf contracts into a well-developed rosette of leaves. The stem hides in the root collar and remains contracted throughout the life of the plant, while the living formative processes shoot up into the spiralling, tightly packed leaves. The leaves lie close to the soil, especially in winter, and in spring they straighten up. The young leaves growing at the centre of the rosette are at first upright and then grow more horizontal and closer to the earth the older they become. As new leaves grow and develop from the spring through to the autumn, the old ones underneath the rosette die off.

The strong and powerful leaf begins with the flexible central rib, which also carries the milky fluid. The leaf blade connected to this runs along the leaf stem and is narrower towards the base. In the leaf sequence, the upper portion is rounded and stretched from the seed leaves onward. But then the expanding impulse of the leaf blade encounters the forces of form and contraction coming from outside. It pushes outward, forming mighty pointed teeth while the counter activity creates great indentations. An image is thus created of strong, partially asymmetric, bizarre-looking, stretched leaves. Only with the helmet-like triangular form at the tips of the leaves does a harmonious balance between growth and form arise.

The flower

The polarity of contraction and expansion or movement already makes the tap root, thickened by the upward contracting pressure of the milk sap, into an almost independent organ of the dandelion. No less independent, on a higher level, is the unfolding of the leaf rosette. As though distancing itself from them, the flower emerges with no transition, removed as it were from what is earth-bound to appear in complete mastery of this polarity.

Towards autumn we find at the heart of the rosette, in the leaf axles, the already-formed flower buds. They overwinter 'under the earth' to the extent that in the autumn they are drawn down into the earth by the contracting root crown. It is a last impulse of movement before winter dormancy sets in. In springtime however, during the April nights, the dandelion develops a kind of vertical stem that shoots up. Using all the pent up forces gathered during the late spring, summer, autumn and winter of the previous year, the flower buds rise up at the top of the milk-bearing, hollow, air-filled, tubular stems. The flower bud appears as a little round green head surrounded by many layers of scale-like bracts. These open one after another in the warm rays of the morning sun and bend downwards. Just a few of them continue to surround the ray-like florets as they unfold.

As the last scales of these bracts drop down, 100–200 golden yellow florets appear closely pressed together on the extensive flower base – a second rosette at a higher level that is turned skywards. They open up gradually from the perimeter towards the centre, the flowers following the movement of the sun. In late afternoon a portion of the bracts bend up into the vertical once again while the flower base simultaneously sinks, causing all the florets to stand encased, bud-like, in a tight bundle. In fine weather this rhythmic opening and closing can continue for several days; when it is wet and overcast the flowers remain closed.

After flowering, some of the bracts raise themselves one more time. They hold all the flowers firmly within them as seed formation proceeds. Meanwhile, the tube-like flower stems grow even taller. During this last stage of envelopment an inversion takes place on the flower base in relation to seed formation in the following way: the seed stands with its head in the flower base while the opposite pole, where the finely divided sepals sit, strives heavenwards. As the seeds ripen, a new movement impulse is activated. At the (upper) sepal pole of the seed, a thin stalk develops that carries the feathery sepals or 'parasol' at its tip. As they expand upwards the withered sepals are pushed up out of the enclosing bracts. After this long period of preparation the bracts open downwards for one last time and the flower base becomes a ball upon which the long-stalked filigree flower, the 'dandelion clock', appears. Each of the single parasols are completely formed since all the seeds have developed in the flower base (receptacle). Now ripened, the seeds loosen themselves from the flower base and with the next breath of wind, sail away on their own.

Diagram 7.1: The different stages of coming to flower, fading and forming the dandelion clock when the seeds are ripe (from Extraordinary Plant Qualities for Biodynamics. Bockemühl, Järvinen, 2006).

The multiple opening and closing of the leaves surrounding the flower, the movement in the flower base and the radiating out of the flower colony into the surroundings, are all expressive gestures that indicate the activity and close proximity of the soul or astral body of this highly developed plant of the *Compositae* family. As if that were not enough, in its process of seed formation it transforms the fine sepals into glistening silver parasols that order themselves like stars in an image of the cosmos. The seed leaves, which normally develop in the moist earthy realm, invert and turn themselves towards the warmth and light of the cosmos. In a further enhancement, there are moments when a light- and air-filled inner space arises within the transparent envelope of the parasol. We could perhaps ask ourselves whether this delicately inscribed inner space is not a true sense-perceptible image of the capacity of the dandelion to draw in silicic acid from the cosmos and truly be 'a messenger of heaven'.

The milk sap

Along with the dandelion, many other members of the *Compositae* family have milk sap in their tissues, including wild chicory (*Cichorium intybus*), milk thistle (*Lactuca serriola*), sow thistle (*Sonchus arvensis*) and the spurge family (*Euphorbiaceae*). It is a milky-white liquid, which in the case of the dandelion – and this is unique – is carried in tubular canals throughout the entire plant. From the thickened, beet-like root they extend through the stem in the root crown and the central rib of the leaf, right up the hollow flower stem to the flower base. The milk sap connects the seemingly separated parts of the dandelion into one whole. We are then faced with a riddle. Should this milk sap not be classed equally as being of a Salt and a Sulphur nature, and even more as a Mercury one (in the sense of Paracelsus' alchemical principles)? Does it not bring together all three qualities? Does the basic principle of a flowering plant, namely the refinement and etherisation of mineral substances, exist here?

The milk sap is a secretion from the lateral cells of the interconnected milk canals that have grown together. It is a substance created by the etheric organisation of this plant, which contains everything in the exact proportions needed to create the highly specialised forms of the tap root, leaf rosette and composite flower. Analysis reveals such a diverse range of dissolved and suspended substances that it seems justified to suggest that in the dandelion we have a primeval form of life from the earliest stages of earth evolution, one in which the mineral, plant and animal kingdoms were not yet separated from one another and still had a form of omnipotent life substance. Certain groups of substances are susceptible to considerable seasonal variation, such as inulin, a polysaccharide formed from fruit sugar, which is almost completely depleted in spring but attains high levels in the autumn. In spring, the root has the most bitter substance, in August the most inulin, in September the most taraxin, in October the most levolin. The milk sap consists of a water-like fluid base – taraxicum liquor – in which many mineral substances including calcium and silicon and organic compounds like proteins, tannins, alkaloids and vitamins are dissolved.

The substances suspended as fine droplets in this solution are resins, mainly gums with colloidal protein shells. In the ash made from the whole plant we find 7% silicic acid, 40% potassium oxide, 8% magnesium oxide, 29% sodium oxide as well as traces of zinc, copper, manganese and sulphur. These figures reveal little about how they relate to the creative process in which they are embedded. More is gained by focusing on certain combinations of substances, such as the active ingredients and their healing effects. But even these say nothing about the whole nature of the dandelion that gave birth to these effects. For as long as this totality was accepted as a given fact of nature, the dandelion was acknowledged to be a medicinal plant. Now, when the active ingredients seen as having medicinal value were isolated and synthesised, or could be replaced by other synthetic substances, the dandelion lost the honour of being a medicinal plant. It will only regain this status – along with all other medicinal plants – when an attempt is made to understand the relationship of the substances to one another and how they have been crafted under the direction of externally working astral forces of the etheric or life body of the particular medicinal plant. The single organic compounds have retained their influence on the etheric organisation as a whole, yet do not fully represent it.

Spiritual scientific research shows that, unlike in the case of yarrow, chamomile and stinging nettle, there is no connection between sulphur and potassium, calcium and iron. There is, however, one between silica and potassium. The living task of potassium is to connect the etheric body with the physical body. Silica, working in a different way to sulphur, forms a kind of sensory connection between these two members and the cosmic-astral forces that work in from the metabolic pole of the farm organism. We can assume that this special interaction of potassium from below and silicic acid from above, comes to fruition through the milk sap that flows throughout the plant. In this context we can see the dandelion in its external threefold form as the 'mediator' between the fine homeopathically distributed silicic acid in the cosmos and what is actually needed as silicic acid throughout the region in which it grows. The same question arises here as arose in connection to the nitrogen of the air: an abundance of silicic acid is found in the soil in solid, colloidal and dissolved form, why then seek to gather such tiny amounts of silicic acid from the cosmos in such a complicated way? It evidently concerns two polar opposite ways in which silicic acid, or the silicon within it, works. The one state in which silicic acid is found is in quartz and silicates. They are the results of growth and decay in past conditions of the earth. As rocks and as products of erosion they form the mineral foundations of our soils. The root of the plant has a connection to this earthbound silicic acid. The other condition of silicic acid applies in the metabolic pole above the earth as 'fine, homeopathically distributed' silicic acid. The immediate thought might be of the meteoric dust that is attracted to the earth. This however comes to the earth of its own accord. This mineral 'rain' is certainly not what is meant. The actual statement of Rudolf Steiner is that 'in the plant there simply must arise a clear and visible interaction between the silicic acid and the potassium.'[5] It is an active process proceeding from the plant and one that is to be brought to life with the help of suitably prepared compost and manure.

What is meant by a cosmic silicic acid indicates a non-material, etheric-astral condition.

If we continue manuring at random from year to year, we *can* gradually prevent the Earth from drawing into itself what it needs by way of silicic acid, lead and mercury ... These influences need to be absorbed into the growth of the plant, if it is really to receive all that it needs from the Earth.[6]

The earth loses the capacity to take up these cosmic substances. In order to counteract this loss the preparation of a special form of manure is needed – the dandelion preparation. It imparts to soil and plant the capacity possessed by the dandelion in a special way, namely, to bring potassium and silicic acid into such an interactive relationship with the living processes of the plant that they are able to 'draw in the cosmic properties'.[7] This interactive relationship only arises in the dandelion because it can lift both the potassium and silicic acid out of the inorganic-physical into an etherised condition via the root.

The kind of silicic acid rayed in from the cosmos and taken up in the living processes of plants connects itself to the silicic acid taken up out of the earth through its interactive relationship with potassium. It is this form of synthesis that gives us a glimmer of understanding for the mysterious statement Steiner makes about the silicon contained in the silicic acid when he says that it is 'transmuted into a substance of great importance, which, however, is not yet included among the chemical elements at all.'[8] Is this the transformed substance that can make the fine substances of the cosmos that are referred to, accessible to plant growth in a new way?

The dandelion unites within itself one of the earliest evolutionary expressions of life, namely the milk sap, with a fully developed contemporary cosmic-earthly formative power. That is what its appearance indicates. Could it be this unique capacity of the dandelion (primarily through its enhanced flowering process) to bring together the silicic acid radiating out from the earth and that radiating in from the cosmos, that transforms the 'silicon in the silicic acid' into a new substance? A positive answer to this can be found in the procedure for making the dandelion and its effect as a finished preparation.

The peritoneum or bovine mesentery

How can the potential forces of the dandelion be captured, concentrated, stored and brought into a form that can be used to benefit the soil and the plants growing on it? With the fading of the dandelion flowers this potential dissipates. The final living stimulus is the transformation of the sepals into the fine stalks that carry the seed parasols. Before it comes to this elaborate conclusion, the living dandelion process that reached its culmination in the flower heads must be kept mobile. This can only be achieved by once more using an organ taken from the next highest kingdom of nature: the animal kingdom. According to the spiritual research of Rudolf Steiner, this task is fulfilled with regard to dandelion flowers by the peritoneum or, more precisely, the bovine mesentery.

The peritoneum, or serous membrane, surrounds the abdominal cavity as well as the pelvic cavity and all the organs contained within them. In the abdominal and pelvic cavities the outer membrane of the true metabolic pole is formed. The peritoneum is an inverted skin inside the body. Its surface, facing inside, forms plate cells that are embedded in a basal membrane. These are attached to the connective and muscular tissue of the organ walls or body cavities. If they are removed, one has a glistening, transparent membrane-like skin in one's hands. It is permeated by a fine network of nerve fibres centred in single ganglia that come together as part of the large ganglionic centre known as the solar plexus. In cattle this is found beneath the spinal column in the transition zone between the breast and lumber region. These nerve networks belong to the vegetative nervous system, which is divided up into sympathetic and parasympathetic sub-systems. Their regulatory activity takes place deep in the unconscious.

Observation of the peritoneum will explain very little of its deeper significance. It can be established that there is a high degree of nerve penetration, a film of moisture that gives the organs in the abdominal cavity, and especially the convolutions of the small intestine, their slippery nature, and an extraordinary capacity to reabsorb bodily fluids. Together with the lymph and lymph nodes it cares for detoxification – in the sense of digestion – of foreign substances. This limited way of observing the more external facts can be extended by questioning why there are so many nerves. The peritoneum provides an answer. It is a sense organ that focuses on what is happening in a highly specific way in each of the organs of this inner space while simultaneously assessing these unconscious processes as a whole. Out of that totality it coordinates the work of the organs in relation to one another. The peritoneum is woven into the intrinsic existence and activity of each of the organs in the abdominal cavity, while also transmitting the sum of these activities to the ganglionic centre of the solar plexus.

There has been some confusion over which part of the peritoneum is suitable for the preparation. In his agriculture course Rudolf Steiner names the bovine mesentery:

Gather the little yellow heads of the dandelion and
let them fade a little. Press them together, sew them
up in bovine mesentery, and lay them in the earth
throughout the winter.[9]

As to what should be understood by bovine mesentery, he
replied: 'The peritoneum. That surely is the generally accepted
meaning.'[10]

Is it the peritoneum or the specific part known as the
mesentery that is to be used? If it is the first, then the greater
omentum (*Omentum majus*) lends itself more readily for the
preparation than the lesser omentum (*Omentum minus*), which
connects the liver and stomach with a double membrane. The
greater omentum sits like an apron between the ventral side of
the stomach and the intestinal mass. It hangs down from the
stomach to the underside of the abdominal cavity, forms a big
loop, turns back up again and cloaks the intestines in a warm
protective covering. It is different however with the mesentery.
It is attached to the wall of the abdominal cavity directly below
the vertebrae and carries the entire intestinal mass. From both
sides of the abdominal cavity the single-layered peritoneum
merges to become a double-layered serous membrane – the
mesentery – with its active sensory surface directed towards the
cavity (see Diagram 7.2). The double serous membrane splits at
the intestine and then surrounds it with a single membrane. In
this way the mesentery forms a wreath of folds, and fills a large
part of the abdominal cavity with the tightly packed intestinal
loops. Between the two serous membranes of the mesentery
are nerve fibres (they branch out on both sides and remain in

close contact with the metabolic activity of the small intestine), arterial and venous blood vessels (they maintain the intense life processes of the intestine), lymph veins (they absorb the digestive juices through the intestinal mucus), lymph nodes (they detoxify) as well as connective tissue and fat deposits. The structure of the mesentery as part of the peritoneum is therefore such that it is simultaneously both a sense organ and an active organ. It transmits metabolic activity (external world) to the whole organism (internal world) and vice versa.

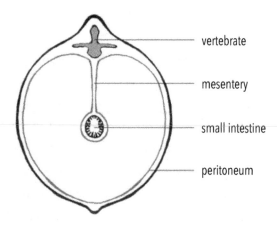

Diagram 7.2: Cross section of the bovine abdominal cavity. The double membrane of the peritoneum in relation to small intestine metabolism.

If we consider this connection with activity, the difference between the mesentery and the omentum (both majus and minus) becomes clear. The concept 'peritoneum' used by Rudolf Steiner in the question-and-answer section of the course with regard to the mesentery, referred to something generic as well as the idea of an all-encompassing, unconscious sense organ deep within the metabolic process. The concept 'mesentery' mentioned in the fifth lecture of the Agriculture Course refers to a specific functional field in which the general becomes a highly specific activity – which is only to be found complete within the digestive tract. The predisposition for it is found in the long journey of intense metabolic activity that culminates in rumination. The result of this opening up of substances via the digestive process is perceived by the serous membranes of the mesentery. It radiates into the body's interior as well as into the veins where it is carried to the nerve-sense pole of the head and up into the horns where it is stopped and rayed back. This releases on a higher level a new consciousness that is again reflected back into the body's cavity and via the two serous membranes imparted to the content of the intestines. These are now permeated by forces which through the current living processes of the cow, have found their way to the supersensible reality of its being. Through the process of excretion this potential enters the outer world. It is what gives cow manure its unique and lasting fertilizing power. This process coming from the cow's higher nature, gives a further indication that the peritoneum presents something 'general' while the mesentery with its two membranes is something specific. This latter fulfils in the 'internal sky', in the inverted cosmos, the core task namely, of imparting the metabolic processes of the ruminant to the organism and from there back to the digestive system to produce manure.

This process involving the entire being of the cow is something that neither the greater nor lesser omentum can fulfil.

Both have specific functions outside of the metabolic process and the transformational steps towards an enhancement of active forces that are connected with it. A characteristic of the greater omentum is the web-like distribution of deposited fatty tissue. This shows that the sensory function is fundamentally different. It is directed towards perceiving and guiding the harmonious working together of the organs in the abdominal cavity, and especially the activity of the small and large intestines in relation to the stomach and the rumen.

As a highly developed member of the *Compositae* family the dandelion has the power to draw in the essence of silicic acid from the cosmos. For its part the bovine mesentery, thanks to its internally focused capacity to perceive the processes of metabolism in the small intestines, has the power to imbue the still immaterial silicic acid with a kind of inner sensitivity. In order to make use of what exists in potential as a result of the evolutionary development of dandelion flowers and the bovine mesentery, a preparation is needed that can bridge the divide between the kingdoms of nature in the seasonal cycle by drawing on human spiritual knowledge.

Stages of the preparation

The best time for picking the flowers (in Central Europe) is on a sunny day in April. The flowers begin to open early in the morning and by ten o'clock are usually fully open. It is best to pick the flowers when the florets in the centre of the flower heads are still closed and pressed tightly together. Pick them without their hollow flower stalks and immediately lay them out to dry. Picking in the afternoon when the flowers are fully open, or late in the flowering season, should be avoided because of their tendency to go to seed.

If they are dry, lightly moisten the flowers with a tea made from dandelion leaves, then press them together a little, surround them with a sheet of mesentery and tie them up with string (not artificial fibre) to make a small package. An alternative to a mesentery, the greater omentum, can also be used; it is easier to handle, is larger and generally less fatty. It is attached to the stomach and otherwise has no direct contact with the intestinal coil.

The envelopment by the mesentery membrane leads once again to an initial inversion of the natural process and hence to a first stage of emancipation. In this instance the outpouring cosmos-inclined gesture of the dandelion flowers is turned inward and fills an inner space, whereas the mesentery, which held all the abdominal organs and contained the loops of the small intestines, is now turned outwards. The double or serous membrane of the mesentery now engages in a new form of sensory activity that is directed both outwards and inwards. Towards the inside the serous membrane is focused on an evolutionary process that has culminated in the dandelion flowers, while the other serous membrane opens itself up to what is streaming to the earth from the cosmos. Seen in this light the once unified sensory function of the bovine mesentery now splits into twin polar functions. The connective tissues of both membranes remain unaffected and maintain the connection between the two opposite spheres of perception.

The second stage of the preparation has gained more practical interest following the publication of Rudolf Steiner's notes to the lectures. Here he noted that the dandelion should be put 'in the mesentery and hung up in the air.'[11] No mention of this is made in the fifth lecture; instead he goes straight to the third step by saying 'we must expose the dandelion to the influences of the earth, and in the winter season.'[12] And that was the practice during the following decades. Yet the question remains open as to whether the second step was dropped or to be understood as described in the sentence that followed: 'Here, too, we must gain the surrounding forces by a similar treatment as in the other cases.'[13] The riddle is solved by the above quote from the notebook about being 'hung up in the air'. The 'surrounding forces' are interpreted as those active above the earth during the summer in the air and warmth. The comment that they should be treated the same 'as in other cases' must therefore have the same meaning as it does in the case of yarrow where its exposure to the summer and winter forces is extensively described.

In the second stage of the preparation another inversion takes place, and with it a further emancipation from what is given by nature (see Diagram 7.3). What was previously part of the cow's internal life, and in whose service it owed its existence, is now an organ of the external world in the service of forces streaming in from the cosmic surroundings. The opposite is true for the dandelion flowers in the world outside: they turn towards the cosmos and owe their existence to it; now they are enclosed in an animal organ where the concentrated forces of the dandelion flowers unite with the forces of the macrocosm mediated by the bovine mesentery.

The flowers gathered in April should be dried and immediately put into a section of mesentery from a middle-aged cow, from one's own farm if possible. The package is then tied together and hung up in the air, protected from birds. During the summer until about the end of September, it is exposed to the solar and planetary forces of warmth and air. These interact with the substances of potassium and silicic acid that have been raised to an etheric condition in the flowers. This is where the connective relationship between potassium and silicic acid comes about on an etheric level, and where, according to spiritual research, the dandelion is given the ability to draw in cosmic silicic acid. This demonstrates why the first stage of the preparation is followed by that of exposing the filled mesenteries to atmospheric forces.

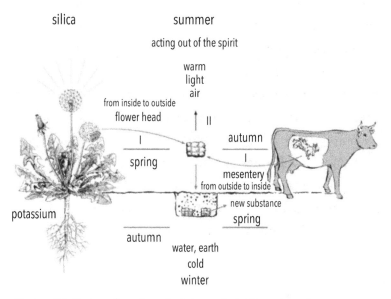

Diagram 7.3: The transformation of substances in the three steps of preparing the dandelion preparation.

In the third stage a further process of inversion and emancipation takes place (see Diagram 7.3). What has spent the summer above the earth exposed to the peripheral forces of the planetary system, is now subjected to the forces coming from the realm of the fixed stars via the earth and water in the soil. During the summer it is mainly astral forces that ray in from the sun and planetary spheres and imprint themselves on the substance of the dandelion flowers through the mediation of the animal sheath, whereas during the winter it is above all the 'I' forces working from higher spiritual worlds that influence the flower substance as it lies buried in the earth.

The procedure for the third stage of preparation is as follows. At the beginning of autumn, around the equinox, the round packages are taken down and buried in the earth in the transition zone between the humus-rich topsoil and the sand- and clay-rich subsoil. The site is filled in again and secured with wire netting against grubbing out by dogs and foxes. Until the mesentery smell has dissipated, a wood or stone covering can serve this purpose too. In spring, around Easter time, the preparation whose mesentery covering is by then strongly decomposed, is taken out and stored in earthenware pots surrounded on all sides by peat. With the finished preparation whose flowers have largely lost their structure, a new substance has been created, or rather a new arrangement of substance that has new kind of potential power.

Effects of the preparation

The dandelion preparation works, as do the other preparations, to create, form and give structure to the fertile soil as well as to the fruit-bearing plants. It integrates them into 'middle realm' of the farm individuality, which corresponds to the diaphragm between the head and the belly.

As a result of the various stages of preparation it has undergone, this new fertilising agent has become

> ... thoroughly saturated with cosmic influences ... It will give the soil the faculty to attract just as much silicic acid from the atmosphere and the cosmos as the plants need, to make them really sentient to all that is at work in their environment. For they of themselves will attract what they need.[14]

It is then explained that these surroundings do not merely extend to the root-filled soil around the single plant, but reach much further. This initially mysterious statement can be explained using current scientific knowledge about the symbiosis of plant roots and mycorrhizal fungi, which act as extensions to the root system and can link plants together over long distances. This network of fungal cells supplies plants with water and mineral salts in exchange for their sugars.

It is to be expected that, as a result of this beneficial symbiosis, the dandelion preparation will be able to make a significant contribution to the soil. Through its link with the earth (via the rising milk sap) and at the same time through its predisposition and openness towards the sun, the dandelion is given the ability to draw silicic acid out of the cosmos. The dandelion preparation is then able, via the compost and manure, to give the plants being cultivated the ability to find the substances that it needs for healthy growth in the soil and wider environment.

With the theoretical scientific concepts of today the idea of such mobility of substance is unthinkable. It becomes more conceivable when the soil itself – the diaphragm of the farm organism – is understood as a living ecosystem into which the root submerges itself as a kind of sense organ. This as yet poorly developed sense organ can be completely blunted by the application of mineral fertilisers. But it can also be trained and become ever more active through the cumulative effect of the preparations we have been referring to here. And that is what it is about. Each and every one of the preparations bears the potential for developing the sensory organisation of the plant and thereby enhancing sensitivity towards substances and forces.

The dandelion preparation is applied in homeopathic doses along with the four previously described preparations, as an addition to compost, manure, liquid manure and slurry. A pinch between three fingers is the amount of each preparation required to inoculate around 1 cubic metre (35 cu ft) of garden compost or between 10–15 cubic metres (350–530 cu ft) of manure or compost windrows, deep litter or slurry. The heap should be prepared immediately after it has been set up and the slurry tank as soon it has started filling up. Instead of suspending portions of the preparations in little bags, they can also be kneaded into clay balls and added in this form to both solid and liquid manures. It is worth repeating this each time the heap is turned or after the liquid manure has been stirred or aerated. The preparations dampen down the re-activated decomposition processes and hence any excessive warming or odours. Nitrogen and carbon remain in stable organic compounds.

As with all the preparations the dandelion calls for spiritual awareness. Only this can awaken questions and encourage an investigative attitude towards the whole farm. A path of knowledge then opens up which lifts the day-to-day management of the farm to a form of artistic activity.

8: Valerian (507)

The valerian preparation is the last of the six compost preparations, mentioned briefly by Steiner, almost as an aside, at the end of the fifth lecture of the Agriculture Course:

> Add this diluted juice of the Valerian flower to the
> manure in very fine proportions. Then you will
> stimulate it to behave in the right way in relation to
> what we call the 'phosphoric' substance.[1]

No further treatment of the valerian juice is required apart from diluting the extract in warm water. When we apply it, it creates a protective sheath for the compost pile, like the animal organs that provide sheaths for the other preparations. It turns the pile into a kind of self-contained totality.

Valerian's characteristics

Valerian (*Valerian officianalis*) is found in transition zones bordering on cultivated land, and grows in loamy, humus-rich soils. It prefers moist, somewhat shady places such as the edge of woodlands, water meadows, on the banks of streams and rivers, and at the base of hillocks where water presses up from below. In mountain valleys it appears at quite high altitudes. It can also be cultivated in the garden. While the dandelion remains hidden among the surrounding grasses and herbs apart from its brief flowering period, valerian rises from its early rosette to stand upright and proud above the surrounding vegetation with its pink-tinged white flower heads. They are made up of closely held together, tiny, delicate single flowers. The stem principle is dominant throughout the plant. Growing to a height of up to 2 metres (6 ft), it reaches down into the root zone where it subdivides to form clumpy rhizomes, and up into the air to form the delicate flower stems. The leaf metamorphosis is most clearly expressed in the first year after sowing. The odd-pinnate leaves grow in alternate pairs and rise from the rosette with ever greater intervals. The leaf stem is gradually reduced and as it transitions to the flower head the leaves shrink right back into the stem. In the following year the leaf metamorphosis is less marked (see Diagram 8.1).

It is striking that just as in the form of the leaves, a metamorphosis takes place on the level of substances. From the root via the stem to the flower the aromatic substances evolve and have their greatest intensity in the flower. While form and colour speak to the soul via the sense of sight, the sense of smell opens up a deeper experience of the activity of substances. The sense of smell leads more deeply into the quality of the astral forces active in the living realm.

With valerian everything that has been formed as substances in the leaf, rhizome and stem through astral influences is allowed to stream out into the air and warmth. It radiates out everything that the formative forces of its associated astral organisation has created in terms of substances. In its openness towards the world, valerian expresses the warmth and light ethers; it is a releaser of warmth and light.

The root

The valerian seedling forms a tap root that soon withers, however. To begin with the stem is held back, and from it a loose rosette of leaves develops. In the earth, the root subdivides into several short, barrel-shaped rhizomes (a form of vegetative fruiting in the root realm) from which root strands grow outwards and then curve down into the depths. They surround a spherical inner space that opens out beneath them from a system of fine roots that is characteristic of the 'earth root' type of plant (see Diagram 8.1). At the transition to the slightly thickened root – which also contains aromatic substances – buds form in the autumn that grow into new shoots. In addition to this form of vegetative propagation, runners are also produced that take root at the nodes.

The most striking characteristic of valerian is its intoxicating scent (or some might say its stink), which is held back and released if the root is cut. Chemical analysis points to the presence of a wide spectrum of carbohydrate substances such as proteins, oils, alkaloids and resins, as well as organic acids and minerals, all of which underpin the many medical indications concerning the calming and healing effect of valerian. All the substance combinations present in valerian are concentrated in its roots and rhizomes. Because of this its medicinal application is focused on complaints of the vegetative nervous system, sleeping problems and nervousness.

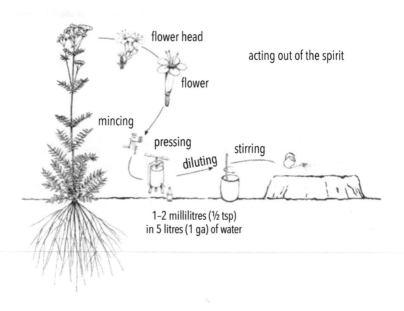

Diagram 8.1: Valerian and its production and application.

The stalk

The stalk of valerian can justifiably be considered as an organ in its own right (see Diagram 8.1). While the dandelion never leaves its rosette stage and has its head of florets in close proximity to the earth, the valerian stalk rises up vertically from its loose rosette in spring. It is as though it seeks the greatest possible distance from its roots and lifts its flowers into the sphere of warmth, light and air. The stalk determines the gesture of this plant and, as its dominant feature, it links the highly active roots with the cosmos-striving flower. Just as it does down below in the roots, it branches out in the air above into the finely divided flower head. The living and supporting tissue of the stalk surrounds an air-filled space, its outer surface is furrowed. As the stalk grows higher the substances are gradually refined. Even the stalk exudes the typical smell of the valerian oil, now refined to the quality found in the flower where the scents of the different essential oils are blended into one.

It is remarkable to experience how the flow of sap in a perennial herb like valerian is formed. The assimilation stream (the phloem) first draws the growth of root and rhizome as well as the substances they contain out of the leaf rosette, which already has the material structure of the whole plant within it. As soon as the stalk starts to shoot upwards these prefigured substances enter the rising xylem stream and are compositionally refined in the air, light and warmth by the formative forces of metamorphosis in the leaf region. The nature of valerian is thus expressed in visible form in the stalk, leaf sequence and flower. The heaviness of the substances in the root is transmuted into 'lightness' and evaporates via the leaves and flowers.

The leaves

As a seedling the valerian goes through all the stages followed by herbaceous plants. The first leaf after germination has a long stalk and an oval-round leaf blade. The succeeding leaves lengthen, and the leaf blades begin to subdivide into irregular, round and on the edges slightly toothed, pinnate leaflets. As the leaf stalk lengthens, their pinnate form grows longer and more pointed. At this point the leaves are upright and form themselves into a bunched rosette. With the stretching of the stalk, the leaf pairs develop one after another in a cruciform way with notably large distances between them (see Diagram 8.1). The pinnate leaves grow progressively smaller and more pointed. As autumn approaches the leaves become shorter, the form denser and the stem more all-encompassing. Finally, as flowering commences, the leaves recede into the stem and only the top leaves remain until these too disappear into the flower. In boggy and shady locations the valerian has a tendency towards rampant growth.

Just as the diversity of substances in the skin of the root forms a relationship with the leaf rosette, so too does the leaf sequence with the flower head. The metamorphosis of the leaves up the stem towards the flower is a visible expression of the invisible formative and transformative processes taking place. In its upward growth the shoot rises from the earthy-watery realm into that of warmth- and light-filled air. This is where the forces of the sun and planets stream in directly. Only now does the archetypal image of the valerian, imprinted in the substances of the rhizome, appear in its totality. Just as the forces active in the watery earth realm have the upper hand in the root, so too in the upward-striving shoot, do the forces of light and warmth mediated by the air. Is this perhaps the reason why the dull, consciousness-lowering smell and bitter taste of the root is in that form virtually absent in the stem, though still present in a less intense way? In the roots these qualities are more intense and concentrated, and as they process through the leaves they become more and more mobile.

In observing and reflecting on the changing form in the leaf sequence, we can shed light upon this dulling of the smell and taste experience. The form of the leaf as it stands objectively before us is what we can take hold of in our thinking. We can see how the leaf that we are looking at is predetermined by the form of the previous leaf and that it will develop further. The transition from one form to the next occurs in the stem segment where one form vanishes and the next appears. Between one leaf form and the next is a connection that remains hidden to physical eyes. The key to recognising this hidden transition is to train the 'thinking eye' on these changing forms. In this way the point-focused, established way of thinking becomes more mobile; self-created thought images are then produced based on illusion-free observation. The same formative forces of the 'creative word' is active in nature and on a higher level also in human beings. If we try to think our way into these phenomenon-based thought pictures, we will be following a path of knowledge in which the hidden connection that links the sequence of leaves on the stem will gradually reveal itself. It is the archetypal image of plant nature that rays out from the depths of the earth towards the sun and unites itself with the etheric and astral formative forces streaming in from the cosmos.

Living in this way into the principle of metamorphosis that is active throughout nature, can be a key towards gaining a comprehensive understanding of the preparation plants, and developing the capacity via their preparation to bring new life to the soil and plants. The metamorphoses take place in the connecting Mercury realm between the substance-transforming Salt pole of the root and the Sulphur pole in the flower. With regard to form it is the other way round – the flower dies into the differentiated forms of its various parts thanks to the activity of terrestrial forces. The root by contrast grows, from its root tip, under the influence of cosmic forces and in the case of tap roots reveals a uniform form. The agency of metamorphosis is the mercurial element that lies in between. Observing this meditatively trains and enlivens our thinking.

The flower

In contrast to the root of valerian, whose strong astrality makes itself felt in the composition of many different substances and especially in the essential oils whose dull and often unpleasant aroma is held back by the roots, the flower releases this scent in a more refined and purified form. Valerian blooms for a long period, from the end of May until September, peaking around June and July. It begins with the straightforward branching of the main stem from the upper leaf nodes. The two-sided branching repeats itself several times so that the main shoot grows beyond the side shoots. This gives the head of flowers a dome-like appearance. At the base of the side stems the last of the pointed, pinnate leaves of the leaf sequence, huddle together. The main stalk and its side branches continue to further divide accompanied still by small pointed top leaves. In the flower head branching is reduced and then limits itself to two side branches. The main stalk contracts to form a terminal flower bud. The side branches grow beyond this and terminate at a new level, also in a flower bud, which is again overtaken on two sides. This process then repeats itself and a third level is formed. The individual flowers – there may be up to 2,000 – come together and create the flower truss. All the processes in the valerian arise from the earth-cosmos polarity in successive metamorphoses and end in the flower by either dissipating in the air as scent or contracting to form seeds.

The seed is positioned upright in the fork of a side stalk. The ovaries are inferior ovaries, which is why the seed leaves appear on the upper end of the seed. They remain without stalks, and when ripe spread their delicate, feathery, leaflet-like wings and allow the wind to carry them away.

The single flower is in itself insignificant, and its pale pink colouration only becomes fully visible in the complete flower head. When the flowers in the upper layers of the truss are in full bloom the lower ones have already wilted and are ripening into seed. This is the reason for the unusually long flowering period. The flowers, with their five variously sized petals whose bases are grown together as tubes at the base, as well as the three anthers, are strongly asymmetrical. As has already been mentioned, the stem or vertical form principle of the valerian, is dominant – an indication that the spiritual archetype of this plant, its true nature, manifests in the vertical earth-sun axis. This correlates further with the strong relationship that valerian has with light and also with warmth, and the high content of flammable substances also point towards this. Warmth is the primal element and bearer of all existence.

The valerian flower gives off a lot of scent when wilted and especially when dried. There are nectaries in all parts of the flower containing various substances of differing consistencies. In liquid form there are several sugars found on the flower base near to the pistils. Others are integrated into the tissue of the pistils, stigma and ovaries. Their aromatic substances sublimate from the solid-liquid or colloidal state into the gaseous. These are essential oils from the strongly sulphur-imbued, earth-root realm that have been refined step by step through the leaf sequence. They are then released as the milder scent of the flowers. The strong astrality at work throughout the entire formative process of the valerian bears a similarity to the soul qualities living in animals. Proof of this is shown by the range of flying insects and beetles that are attracted to the plant. It is correct to say that the out-streaming scent of the flowers is the purest expression of the nature and activity of formative forces. In it, material substance is de-materialised to the highest degree.

The substances in the flower are to a great extent highly complex compounds of carbon and hydrogen. Oxygen, which brings a greater connection to the earth, remains in the background. That is why these substances are highly flammable and leave no ash behind – carbon takes on a gaseous condition as carbon dioxide. The phenomenon of flammability points towards the fact that hydrogen is the bearer of the element of warmth or its supersensible equivalent, warmth ether. Rudolf Steiner characterised hydrogen as

the substance which is so near the spiritual on the one
hand and to the substantial on the other ... It carries
out again into the far spaces of the universe all that is
formed, and alive, and astral ... It is hydrogen which
dissolves everything away.[2]

That hydrogen has the capacity for dissolution in the
higher sense of the word, makes sense morphologically in
that as flowering commences, the leaves seem to disappear
into the stem and physiologically in that the phase of protein
development is replaced by a de-vitalizing phase in which
hydrogen-rich carbohydrates such as fruit sugars and essential
oils are formed. In valerian, the fructifying sulphur impulse
takes place most intensely at the opposite pole, namely in the
root. The aroma is 'imprisoned' by the root cortex. If the root
is cut into, a thick and penetrating earthy smell is released.
In the flower by contrast, especially when it begins to wilt,
it becomes light and airy, has a flower-like smell and fills the
immediate surroundings with a cloud of perfume.

Imaginatively speaking, the hydrogen process dissolves all
the defined form of the substance into the indistinguishable
chaos of the cosmos, into the original warmth condition.
The formation and dissolution of substances in the valerian,
dominated by the stem principle, is permeated by a strong
astrality. This fact and its special relationship to warmth can
explain why the valerian flowers need no further preparation
with an animal organ, nor do they need exposure to the forces
of the cosmos and the seasonal cycle of the earth. It is also
not possible to find an animal organ appropriate to this plant.

The nature of valerian is in this regard polar opposite to the stinging nettle. While the 'inner astral activity' of the nettle forms new substances within its own containing organ along the incarnating path of hydrogen, in the valerian flowers the opposite, excarnating processes of hydrogen take place. Here, all structure is dissolved into the primal state of warmth whose nature is simultaneously both sheath and substance. The valerian manifests an 'outer astral effect' which as it streams out, creates inner holding spaces.

Gunter Gebhard, a Waldorf teacher and geologist, expands on this:

In the stinging nettle it is sulphur and iron that
are active. Both have an inward centring quality.
The activity of the stinging nettle demonstrates a
connection to the quality of reason. The terrestrial
element is expressed most strongly in the rhizomes of
the stinging nettle. The flower of the valerian dissolves
substance out into the cosmos; it strives towards the
light of the cosmos. It is exactly the same with the
valerian roots that 'radiate' into the earth like the light
in straight lines, without branching and with virtually
no secondary thickening. In the blossom it is the
activity of hydrogen, in the root the activity of light
that reveals itself in phosphorous. In contrast to the
stinging nettle which relates to the microcosmic 'I', the
valerian is connected with the macrocosmic 'I'.[3]

Stages of the preparation

The valerian flowers are picked when the greatest number of flowers have opened in June or July and then immediately pressed. This releases the juice, including the essential oils, from the plant cells. It is then separated from the fibres by being pressed through a juicer. The fibres are then covered with the same amount of water as was extracted as juice in the first process. The mixture is left for a day and then pressed again. By pressing it a second time around 1 litre of juice can be obtained from 2–3 kilograms (1 qt from 4–7 lb) of flowers. It is put into bottles where it undergoes a lactic acid fermentation. The gases created through this process must be allowed to exit. After about six weeks of fermentation, the bottles are stoppered with corks or rubber seals. Too much oxygen leads to decomposition, rendering it unusable as a preparation. The bottles are kept in a dark place with an even temperature – the preparation store. The juice keeps for many years.

With regard to its application, Rudolf Steiner recommended that the flower extract should be 'very highly' diluted in warm water.[4] This rather imprecise statement has been variously interpreted and tested. Our practice is to add just so many drops of the extract to a given amount of warm water as releases the valerian scent and gives the fluid a brownish tinge after about 5–10 minutes of stirring. Regarding quantity, the equivalent to the 'pinch' of each solid preparations is as follows: 1–3 grams (1–2 cc) in 5–8 litres (1–2 gal) of water per 8 cubic metres (35 cu ft) of compost material. The liquid is sprayed over the heap after it has been set up, or poured in portions into holes along the upper edge. With liquid manures (for example, slurry) the diluted juice is best added when the tank is being stirred. It is worth emphasising that accompanying the experience with an open and inquiring mind, is the best advisor.

The connection between valerian and the element of warmth has also found a practical application in biodynamic agriculture apart from its use as a compost preparation, namely when the danger of frost threatens in May. Immediately beforehand, a fine mist of the valerian extract diluted in water at 1 millilitre per litre (¼ tsp per qt) can lay a blanket of warmth over the fruit blossoms or endangered vegetable plants that can prevent damage from up to 3°C of frost (27°F). It is also recommended in vegetable production for strengthening the plants and is sprayed in a dilution of 10–30 drops of valerian preparation in 3 litres (3 qt) of water. It is also strongly prized in viticulture.

Effects of sulphur and phosphorous

The series of six compost preparations were introduced by referring to sulphur in relation to yarrow, chamomile and stinging nettle. It concludes with Steiner's brief, enigmatic statement pointing towards the relationship of valerian to phosphorous. These two elements, sulphur and phosphorous, are very closely related to one another. They are both expressions of light and warmth, and yet they are different substances. Although they are adjacent non-metals in the periodic table, they act as polarities in the organic realm. In the inorganic realm, sulphur is found in the form of numerous sulphides, the metallic salts of sulphuric acid, and the pure sulphur within volcanic deposits. In this form it burns with a blue flame to sulphur dioxide at around 250°C (480°F). Phosphorous is different; it is very reactive and is therefore not found naturally in its pure state. The amazing phenomenon of elemental phosphorous is that it appears in five different temperature-dependent forms. In a first modification it appears as white phosphorous. This is poisonous, glows in the dark and can only be maintained under water. It spontaneously combusts on contact with air at 44°C (111°F) and burns with a bright, glowing hot flame to form phosphorous pentoxide (P_2O_5). With the exclusion of air and a higher temperature, or else in full light, a second modification occurs: red phosphorous. This is stable, does not glow and is not poisonous. There are three further modifications with significantly different properties: pale red, black and black amorphous phosphorous.

Phosphates are to be found in sedimentary rocks and above all in crystalline igneous rocks finely distributed as insoluble apatite – calcium phosphate with traces of fluorine and chlorine. Larger deposits of apatite are mined on the Kola peninsula of north-west Russia. Apatite is broken down by acids and in that way regenerates phosphate in the soil. Phosphorite is found in sedimentary deposits, primarily in north Africa. These are formed of compressed animal bones from the Tertiary period. It can be milled and applied as a slow-release, soft, ground phosphate to phosphate-deficient soils.

The phosphorous content in the soil profile declines from the living topsoil down into the sub-soil. This demonstrates that losses due to leaching are minimal, generally less than 300 grams per hectare (4 oz per acre). Far higher losses are incurred through surface erosion.

The stability of soluble phosphorous (HPO_4^{2-}) formed by the breakdown of minerals and organic materials, is due to its reactivity. It readily forms new compounds, especially with calcium with which it forms secondary calcium phosphates. It is also absorbed into aluminium and iron hydroxide compounds, which in turn form complex humic and fulvic acid compounds. It is found suspended in soil colloids and attached to clay minerals and humus. Up to 80% of the organic phosphorous compounds appear in the form of phytate (calcium and magnesium salts of inositol polyphosphate). A further 5% to 10% is found in the nucleic acids of the cell nuclei of plants and micro-organisms.

It is noticeable how phosphorous is most finely and exactly distributed throughout all the kingdoms of nature and in the human body. If we limit our gaze to the way it manifests physically and materially, very little is revealed of its true being and function in nature. It is very different if we look towards the plant world and beyond. Phosphorous, now transformed into a phosphorous process, is found everywhere in metamorphoses where there is a tendency towards devitalisation. It is directly revealed to our observation in the flowering of the plant, which is essentially a phosphorous process, and is particularly active where devitalisation reaches its culmination in the wilting of the flower and the formation of seeds.

Devitalisation means dying back and becoming pure form. This is physiologically what phosphorous is doing. It transmits the quality of form to the stream of inheritance via the nucleic acids.

Deeply hidden within the processes of growth, there are as many devitalisation zones in the plant as there are cells. These are the phosphorous-containing cell nuclei. In the form of nucleic protein, nucleotides and chromosomes they host the mediator of the stream of inheritance that passes from one generation to the next. The nucleus, which is surrounded by the nucleic membrane, represents the still pole of the cell in relation to the cell plasma. Here, on the basis of phosphorous, a higher principle is at work than that which creates the existing form of plant, animal and human being. It is the principle whose ongoing vital power leads development from the past into the present and from there into the future.

The plant as a purely living being has no incarnated soul and independent self and hence no nervous system, which is the earthly foundation for the astral body. Yet the question arises with regard to the nucleus of the plant cell whether this might not be an evolutionary precursor. The nucleic acids RNA and DNA point towards a higher phosphorous content in the phosphoric acid residues of the nucleotides. Just as the finely branching nervous system in animals and humans extends into the living tissue, we find in the individual cell plasma outside the nucleus, organelles – ribosomes and mitochondria[5] – that relate to the nucleus and plasma as tiny, structured corpuscles. Like the nucleus, the mitochondria have chromosomal structures that we may presume have an epigenetic function, transmitting influences from the plant's environment to the cytoplasm. This includes the effects of the earth, sun and planetary constellations, as well as the cultivation, fertilising and breeding measures carried out by human beings. The latter are recently gained contemporary characteristics, and we must reckon with the fact that such epigenetic influences occur each season. In order to retain the fruit of these terrestrial-cosmic influences during the year (among them being the effects of the preparations), the saving, maintaining and breeding of the farm's own seed varieties is recommended.

The earth-sun axis into which flowering plants are membered is manifested in the stem, stalk, haulm, shoot, branch and trunk. The same power that is active in this upright gesture, which is a reflection of its spirit nature, is found at a higher level in the human being as self-directed will activity and awakening self-awareness. The physical starting point for making these forces effective is phosphorous, not on its own but combined with other substances such as occurs in the nucleotides of the cell nucleus. In relation to human beings Rudolf Steiner explained that:

> It is very remarkable that the human 'I' – if we now regard its spiritual, psychological, organic and also mineralising action in the human being – is a kind of phosphorous bearer ... To bear phosphorous through the human organism, permeating us with phosphorous, is an 'I' activity.[6]

The phosphorous process in its most labile form of adenosine triphosphoric acid (ATP) primarily drives carbohydrate metabolism, the nucleic acids connected with nucleic proteins, the metabolism of protein in the form of phosphatides, the metabolism of fat, and finally the mineralising process of bone formation with the phosphorous-containing calcium or apatite. Different phosphorous compounds give impulses to the various ethers relating to the human members: the 'I' lives in connection with phosphorous in the warmth ether through which it imprints itself through sodium phosphate on the astral body; through magnesium phosphate and the light ether, as well as potassium phosphate and the chemical ether, it works on the etheric body, and on the basis of calcium phosphate and the life ether it congeals in the solid form of the bony system. Through its direct connection to the core spiritual being of the 'I', phosphorous helps to overcome one-sided qualities in the physiological process and balances out contradictions.

The phosphorous process is concentrated in the nucleotides in the centre of the cell. They are the bearers of inheritance and carry the achievements of evolution into the present. All that once arose from the spirit is drawn together into a germ through which it can then receive new impulses to carry into the future to be born again. Through the alternating rhythm of day and night this impulse is taken up by the plasma of the plant cell; it radiates from the light of the sun and is modified by the planets. The place where this germ-like quality is realised is in the protoplasm that surrounds the cell nucleus and contains the previously mentioned organelles. The basic substance of protoplasm is cytoplasm, a clear, transparent, colloidal fluid, and a homogenous mass of protein.

Cytoplasm flows around the cell vacuoles or, in certain situations, around the cell nucleus. It is sensitive to external stimulation. Its cell tissue, for example, can expand or contract in response to weather conditions, and it can cause the opening and closing of flowers or the movement that allows flower heads to follow the sun. It can also respond to the kind of fertiliser used and to cultivation methods. In the same way as a stamp impresses itself on a seal, so do the astral forces pushing in from outside impress themselves on the cytoplasm. Botanical science ascribes a 'sensory capacity' to the cell plasma, but this unique plant quality is not considered a transitional stage towards animal sensation.

Phosphorous dominates the cell nucleus and determines its strict form. Like all protein, the nucleic protein associated with it contains no phosphorous, because protein is associated with sulphur. Protein's substance-forming activity is due to the sulphur that is so closely related to phosphorous, and yet in its properties is so polar opposite. This is, however, the mediator between the astral forces streaming in from the cosmos and the substances that make up protein (carbon, oxygen, nitrogen, hydrogen). Both sulphur and phosphorous serve to harmonise the human members. If, for example, the astral body accompanied by the 'I' enters too deeply into the etheric and physical organisation, then sulphur working with the astral body and phosphorous more with the 'I' will loosen this connection.[7] Both act as the alchemical Mercury process between higher spiritual worlds and their earthly reflection, between the spirit germ-like quality and what is expressed as a finished work in its earthly manifestation. Both appear highly rarefied in all living processes and yet in a polar opposite way, bringing the past into the present and being open towards the future. They are both woven into the cell in highly diluted form – phosphorous in the nucleus, sulphur in the plasma. Their activity could be described as an example of nature's homeopathy, the effect of the smallest entities.

With regard to hydrogen and its relationship to phosphorous and sulphur, it can be summed up as follows. Hydrogen can be characterised as a representative of the fire element or warmth ether. In hydrogen, warmth and light are still interwoven. During the evolutionary journey towards the earth and increasing materialisation, the warmth ether solidified to sulphur and the light ether to phosphorous. Sulphur brings mobility and growth to the realm of substance through warmth (no sulphur, no metabolism), while phosphorous creates form and, on the level of human consciousness, enables the image of what has been formed as substance to be thought. Light is experienced in the soul as thought. The key relationships of each are – sulphur and warmth/movement, phosphorous and light/form. Sulphur, however, with its brilliant yellow flame, also reveals its connection to light, and phosphorous, with its highly flammable nature, its connection to warmth. In living organisms, sulphur rules the metabolism and phosphorous the line of heredity through DNA and RNA nucleic acids. The phosphorous process in the valerian flowers is refined to such an extent that we may assume that through the phosphorous-imbued nucleotides of the ribosomes and mitochondria of the plasma, it has the potential capacity to draw on surplus germ-like forces to initiate new possibilities in plant development.

Effects of the preparation

It would certainly be wrong to see the valerian preparation as a phosphate fertiliser. Valerian contains no more phosphate in its ash than any other plant. What determines its effect are the formative forces concentrated in the valerian flowers. These are able to so structure living processes that they can mediate phosphorous forces from the realm of pure spirit and help the plants attain an upright quality and a distinct existence as an organism, or, in the words of Rudolf Steiner, stimulate the manure 'to behave in the right way in relation to what we call the "phosphoric" substance.'[8] This stimulation is needed wherever organic materials are decomposing, and applies to whenever manure and compost is prepared. Through autolytic decomposition and the microbial breakdown of plant and animal residues, the compositional order held together by phosphorous falls apart. In the resulting chaos the substances, now robbed of their carriers, have the tendency to fall out of life into an inorganic state. Nature herself is to a certain degree able to intercept this dying process by forming stable humus. This process takes place in the moist and watery realm influenced by moon forces. The compact accumulation of organic materials in manure and compost heaps, as well as in liquid manure and slurry tanks, requires human guidance and the use of the measures referred to.

The effect of the valerian preparation, however, goes beyond this. Its effect is not exhausted with the re-integration of free phosphorous into the nucleic proteins of microbes, and fixing it until such times as it can be taken up by the crops. The microbial processes occur in the first place in the moist-watery and not the solid-earth environment. This last only occurs through the activity of the astral bodies of specialised soil animals. The effect of the valerian flower juice in comparison seems to be a universal all-encompassing and unifying process – one which can contain the rampant metabolic activity of the compost heap and turn it into a self-contained totality. Under the influence of this preparation the compost pile develops into a kind of organism in which the chaos of decomposition is penetrated by astral forces streaming in from the periphery. These forces give the compost pile an astral organisation that, by virtue of its indwelling spiritual 'I' nature, makes phosphorous, now released from its former compounds, its carrier once again. Through the addition of the valerian preparation, phosphorous, having divested itself of its carrier role in the chaos of decomposition, attains a new kind of aptitude.

Seen in this light the valerian preparation works via the compost in a most intimate way on the formative processes of life which take place 'horizontally' in the diaphragm of the soil and spread out across the surface of plant leaves, as well as 'vertically' through what is expressed in the power of the plant stem in the spiritual qualities of the heights and depths. Against the background of this 'world cross' we will need to find the observable elements that confirm what is manifested through spiritual research. This applies to all the preparations.

9: Equisetum Tea

The equisetum preparation does not belong to the series of compost preparations. Instead it stands alone and is described in connection with fungal disease prevention by Rudolf Steiner in the sixth lecture of the Agriculture Course. It is the common or meadow horsetail (*Equisetum arvense*) that is used. Along with other kinds of equisetum, club mosses and ferns, it represents an archetypal, primeval plant form. Forerunners like the psilophytes are traced back to the end of the Silurian and beginning of the Devonian periods, which is around the middle of the Paleozoic age. This is the time of Lemuria, the stage in earth evolution when the previous planetary condition of Ancient Moon is being repeated.[1] These primeval types originate in the water and then take to dry land. This proximity to the element of water can be seen in the related flora of that time and in the still-existing equisetum family of today. This is expressed both by their preferred habitat and the complicated germination process. Horsetail grows on loamy and clay soils that tend to be waterlogged or have hard pans.

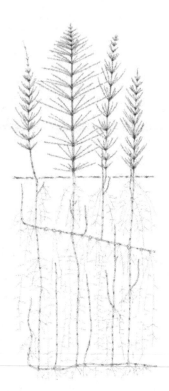

Diagram 9.1: Horsetail with shoots and underground rooted rhizome system.

Equisetum's characteristics

The external appearance of horsetail is surprising in relation to the vegetation that usually surrounds it. The stem, subdivided into many single stem segments (or internodes) nestled into one another, rises straight as an arrow (see Diagram 9.1).

The whorled, undivided side branches follow the same structure. They ray out upwards at a slight angle from the horizontal. Stem and side branches are a uniform green

140

colour and replace the leaves in their function. What is unusual are the missing leaves and flowers. Because of this there is no perceptible metamorphosis of form apart from the fact that the side branches at the base of the stem are shorter, then extend out further, and towards the growing point contract in pyramidal fashion. Leaf development is limited to the elongated and pointed nodal sheaths that lie tightly connected to the base of the stem segments. Their hexagonal box-like arrangement reflects the silica-quartz quality that permeates the entire plant, as does the hexagonal covering of the spore-bearing shoots. The entire plant is dominated by the stem principle. By contrast, fern, though temporally of the same origin, is dominated by the leaf principle; club mosses have an intermediate position. What is so amazing is the contradiction the horsetail presents. On the one hand it is close to the watery moistness beneath the soil, and on the other it integrates its strongly formed silica structure into the vertical sun-earth axis. The horsetail plant masters this contradiction. If we look for the plant's generative pole we will find it separated from its vegetative part as a reddish-brown spindle-like cone, each one sitting upon a shoot with closely pressed together stem branches; at the base they are surrounded by a wreath of brownish-black pointed leaves. The cones are reminiscent of fungi that emerge out of the darkness of the earth. This is what makes the field or common horsetail unique within the equisetum family, and also indicates why it is effective against fungal growth. By separating the fertile spore-bearing cone from the sterile green shoots, it is distancing itself to some degree from the fungal gesture. The other horsetail species all carry the spore-bearing strobilus at the tip of the green plant.[2]

The vertical vegetative shoot extends down with the same segmentation into the depths of the earth (see Diagram 9.1). If it meets a hard pan, for example the boundary between the topsoil and the subsoil or the ground water, it will branch out horizontally and follow the boundary. The rhizome shoots are air-seeking. Secondary shoots from the rhizome nodes and also from the primary vertical shoot, grow upwards towards the light to increase the vegetative plant stock. The vertical rhizome roots break through the compacted zones of the soil with a great deal of force in order then to branch out horizontally along a deeper, water-bearing layer. Even at depths of three or four meters it can send up its brownish-black rhizome shoots.

The horsetail has no primary roots. Its fine, feathery, downward-directed roots come from the shoots, emerging from between the segments, the internodes of the rhizome shoots. In late autumn they develop egg-like swellings on the horizontal rhizomes growing near the surface. These provide a source of nourishment for the first shoots in springtime and especially for the reddish-brown spore-bearing shoots. Such a shoot with a spindle-like form carries a hexagonal spore pocket. When the spores are ripe, the fertile shoot rapidly dies away. The spores, which are as fine as dust, are then carried away by the wind and germinate as soon as they reach a moist and watery environment. The first thing that emerges from the germinating spore is a root-like growing point. It then develops into a green, algae-like prothallium, a heart-shaped structure that produces both male and female sex cells, the former developing into a round-shaped spermatozoa that can move freely in the water with the help of two flagella. The latter forms ovaries that are then fertilised by the spermatozoa. After fertilisation the vertically up-and-down-growing shoot of the horsetail leaves its watery phase and connects with the solid earth and the forces of the sun. Both the fertile and vegetative shoots develop from the rooted rhizomes that penetrate deep into the earth.

The air-, light- and warmth-striving shoot of the horsetail is made up entirely of stem. True leaves are absent. There are only rudimentary leaf forms that appear as insignificant, collar-like protective scabbards connected as whorls surrounding the base of the stem segments. The ribbed segments of the central shoot and the generous ramification of the side branches growing out from the whorls, present a large surface area for absorbing sunlight and giving off water vapour. This is configured in such a way as to ensure the passage of large amounts of water through the horsetail. Its close relationship with water also shows its worth in that it extracts the silicic acid from the upward flowing xylem stream that comes from the weathering of minerals in the soil. Its inorganic nature is enlivened by the life-giving sun forces in the plant. It could be said that in this plant, which existed in an ancient period of plant evolution, the forces of the moon mediated by water make the silicic acid once more receptive to the present day activity of the sun. It is given forces that can create a harmonious balance between the lunar activity of the past and the sun activity of the present. Horsetail is in this way a polar opposite of dandelion. If dandelion as a member of the *Compositae* family belongs to the most highly developed plants, horsetail is one that appeared at the very beginning of the evolution of land plants. In horsetail, the silicic acid of past stages of evolution is brought to the present, in dandelion the present evolutionary phase is led towards the future.

The life process of the horsetail is a water-born, metabolically active silica process. It moves from within towards the periphery of the living tissue. As the water evaporates in the surrounding warmth, the silicic acid is excreted and coats the entire plant with water-holding, amorphous silicic acid. It is present in the epidermic cells and in layers between the cell membranes. It goes through all the states from a solution, via a mobile gel to a hardened glass-like opal – water-containing silicon dioxide ($SiO_2 + H_2O$). If the green part of the horsetail is burnt to ash, what remains after the black carbon structure has gone is a white skeleton of silica. It has lens-like swellings, which in the living state serve to guide the sunlight to the ordered rows of chlorophyll. The high silica content – 70% in the ash – creates the stiff, ray-like form of the horsetail, making it rough and brittle.

Stages of the preparation

The field horsetail is collected on a sunny day around St John's time when it is fully grown, spread out thinly in a shady, well ventilated place and dried. With regard to its further treatment, Rudolf Steiner said:

we now prepare a kind of tea or decoction – a pretty concentrated decoction of *Equisetem arvense*. This we dilute, and sprinkle it as liquid manure over the fields, wherever we need it – wherever we want to combat rust or similar plant diseases.[3]

Walter Stappung has documented the many different ways in which equisetum tea is prepared across the world, the length of time it is boiled and the subsequent fermentation to a liquid manure, as well as the cold fermentation of the fresh plant material, application times, quantity and frequency and whether the preparation only works preventatively or can treat an acute fungal attack.[4] Out of the accumulated experiences of making the tea the following recipe can be given: 200–300 grams (7–10 oz) of dried herb is gently boiled in 10–20 litres (2½–5 gal) of water for one hour. With fresh plant material, 1½ kilograms (3 lb) is added to the same amount of water.[5] A long period of simmering is recommended in order to break open the silica-coated peripheral cell tissues and keep this process in the universal and primal element of warmth. With regard to application Rudolf Steiner uses the concept 'liquid manure'. This indicates that after it has been made, the horsetail tea should be fermented so that the heavier organic substances that are harder to break down can be extracted. With the formation of acids the tea becomes a liquid manure and, as a result, can be kept and used over a longer period. Only when it begins to putrefy and starts to stink of hydrogen sulphide (H_2S), should it no longer be used. The frequently practised cold fermentation should also be avoided. With longer periods of fermentation, premature rotting and an insufficient opening up of the amorphous silicic acid can occur. The effect of warmth in the boiling process seems to be of key significance in making the silicic acid effective.[6]

The application times depend on whether the liquid manure is to be used preventatively or for an acute case. Rudolf Steiner's indications suggest that it is primarily for prophylactic use. The fermented tea works on the soil and reduces the excessively moist and watery conditions that are conducive of fungal growth on plants. From this point of view it is worth applying the fermented tea over the entire farm or garden if possible on an annual basis at the rate of 100 litres per hectare (26 gal per 2½ acres). It would be ideal to do this three or four times a year, during late winter to early spring (March), and in summer and autumn through to early winter (November). In the context of crop rotation it should in any case be applied to susceptible crops, like grains and certain root crops. A direct spraying into acutely affected crops, often in combination with stinging nettle ferment, can have a significant effect. This is a measure that treats the problem. The preventative treatment of the soil helps to bolster resistance.

Effects of the preparation

Applied as a preventative, the horsetail preparation becomes effective above the earth. Following a moist autumn and a mild, wet winter leading into the spring, the entire root zone is very strongly influenced by the forces of the moon, especially so at full moon and perigee. They are mediated by the water and bring about a form of 'lunar vitality' in the soil not dissimilar to the watery life that the lower plants and animals once developed in primeval times. In our day this form of vitality lives on in the world's oceans and, in a different way, in rivers, lakes and ponds. This early, less differentiated form of life, originated in the early to mid-Lemurian epoch, which was a repetition of the Ancient Moon condition that preceded earth evolution.[7] Geologically speaking, this is the Paleozoic period. At that time when the moon forces predominated, there were as yet no flowering plants. It was the age of spore-bearing plants to which fungi belong, as well as the plants already mentioned. The changing conditions of the soil as it developed out of the watery state, saw the relationship between the moon and sun forces change. In the following Cenozoic or Tertiary period (during the time of Atlantis), the sun forces gained the upper hand and flowering plants with their primary roots began to appear. In the Quaternary or Recent period, humus and mineralised soils developed and, with human assistance, became cultivated soils.

If moon forces become dominant during a lengthy wet period, our cultivated soils will be permeated by lunar vitality in a one-sided way. The lower plant and animal forms from that earlier stage of evolution, especially fungal life, will then be stimulated. This occurs not only in the soil but also higher up in the plant. What rightly belongs in the darkness of the soil, rises to form a kind of second soil layer in the air where microbial life forms survive parasitically on the living tissue of plants. Even the power of the sun can only partially limit this rampant, abnormal life. A plant needs to be found that concentrates and processes so much sun power in itself that it can render the excess moon forces carried in the water harmless, and create a healthy relationship between the polarities of moon astrality working from the past and the present day sun astrality shining in from the cosmos. Horsetail has this capacity as an outstanding representative of a plant that bridges the transition between a water-born and an earth-born plant. The effect of the tea ferment is to free the soil of its surplus moon forces. The horsetail preparation

> deprives the water of its mediating power, so as to lend the earth more 'earthiness' and prevent it from absorbing the excessive Moon-influences through the water it contains.[8]

This description clearly points towards the prophylactic treatment of the soil with the horsetail preparation. Living with the changing weather patterns of the year trains us to act wisely and with foresight. A regular prophylactic treatment each year will prevent acute cases from arising. A lot of experiments have been made, primarily in horticulture, that focus on reducing acute cases and other fungal disasters with the horsetail preparation.[9] However, there are no long-term experimental trials regarding the prophylactic effect of the horsetail preparation. Trials of an exact nature exceed what is possible. Being certain about the effects, and this applies to all the preparations, depends on penetrating the results of spiritual science with our thinking and understanding as well as the confirmation of practical experience.

10: How the Six Compost Preparations Work Together

Some might question whether the sequence in which Rudolf Steiner described the six compost preparations in the Agriculture Course was random or significant. Everything tends to suggest that the latter is the case, if we consider the special qualities of the materials used as well as the effectiveness of the preparations themselves in successively building up the 'diaphragm' of the soil between the cosmic heights and earthly depths, and in creating a living, ensouled and spiritual totality.

The preparation plants

The six preparation plants are characterised in such a way that between the 'flower/Sulphur pole' and the 'root/Salt pole' very specific relationships of cosmic and terrestrial substances prevail. It is also striking that the first three preparation plants of yarrow, chamomile and stinging nettle, form a trio, likewise the oak bark, dandelion and valerian (see Diagram 10.1). Between them is a hidden, and at the same time objective, transition to a middle region, which is the soil. The two groups of three form a polarity. The first, the terrestrial pole, is more basic and alkaline. Sulphur is master here. In the inorganic realm it works as an acid; in the organic realm it reveals its cosmic nature in that 'it is along the paths of sulphur that the spiritual works into the physical domain of nature.'[1]

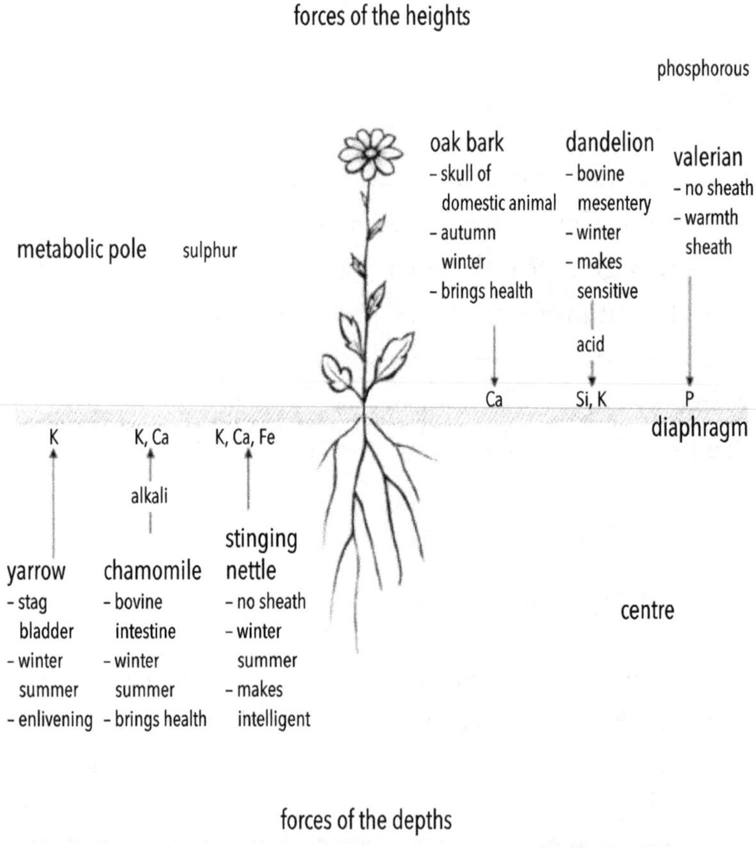

forces of the heights

phosphorous

metabolic pole   sulphur

	oak bark	dandelion	valerian
	– skull of	– bovine	– no sheath
	domestic animal	mesentery	– warmth
	– autumn	– winter	sheath
	winter	– makes	
	– brings health	sensitive	

acid

Ca   Si, K   P

diaphragm

K	K, Ca	K, Ca, Fe
	alkali	
		stinging
yarrow	chamomile	nettle
– stag	– bovine	– no sheath
bladder	intestine	– winter
– winter	– winter	summer
summer	summer	– makes
– enlivening	– brings health	intelligent

centre

forces of the depths

Diagram 10.1: The series of preparation plants and animal organs in relation to the metabolic and nerve-sense system of the animal, as well as to the middle realm created between the forces of the cosmic heights and the earthly depths.

It dematerialises the alkaline metals (K, Na) and the earth metals (Ca, Mg) in the life processes of the plants, as well as iron as a heavy base metal. In the first series of three preparations it is the yarrow with its special sulphur power that processes the alkaline metal potassium. With the chamomile this process goes further and includes the earth metal calcium. Alongside potassium and calcium the stinging nettle then uses its sulphur power to process iron into iron radiations. Iron in nature, and in the human being, is the metal of incarnation: it enables the spiritual element to work into the physical body. In the human being, iron is concentrated in the blood, which connects the heart with the periphery and bodily functions as it circulates. In the blood stream that flows to the heart and then returns to the rest of the body, the soul and spirit of the human being experiences itself as being incarnated in the body.

The heart reflects on and brings together what has been impressed upon the blood in the bodily organs and periphery of the organism. In the heart, the inflowing and outflowing blood has its own active sheath, which, through its rhythmic pulse, keeps the blood alive. At the opposite pole are the kidneys, which in a different and opposite way also maintain the blood's capacity for life. They form a space in which, as an arterial stream, the blood leaves the vascular system and, raised beyond terrestrial-material processes, is sifted and refreshed by the kidneys and brought into a harmonious balance between the terrestrial-material and cosmic-spiritual processes. Only after being 'tested by heart and kidneys' is the blood truly viable.

It is different in the case of higher plants. They grow in a very specific way, as the preparation plants demonstrate, out of the fertile soil – the centre or middle realm between the heights and the depths. This centre is initially there only as potential; it is not yet able to act independently in terms of silica, calcium, clay and humus; it needs cultivating. This applies above all to the functions of clay, which in its seasonal rhythm has a potential heart function, and to humus whose dynamic quality is closer to the function of the kidneys. The measures needed for this cultivation are part of the true art of farming.

Of the many measures that have been characterised, it is fertilisation by the human spirit that opens the door to the future. This form of cultivation concerns the enlivening of substances from the depths. The first group of three – the yarrow, chamomile and stinging nettle, which gradually help the centre, the diaphragm, become functionally independent – can be used for this purpose. These three medicinal plants have the power in their living processes to transform potassium and calcium into nitrogen. This means to enable them to become carriers of astral forces.

With regard to this first group of three, Gunter Gebhard commented:

> Sulphur, as the substance that fell directly from the
> classical element of fire into the element of earth,
> is the substance that leads the spirit into earthly
> existence; in short, sulphur is in essence something
> that can materially condense warmth, which means
> that wherever this essence is active sulphur appears as

substance. The first three preparations are those which promote the incarnation of something spiritual (plant life). In the case of yarrow it is potassium, which, bearing the fluid element wholly within itself, can bring about the etheric connection with the individual etheric body. The calcium with the chamomile draws the astral into the physical body that is permeated by the etheric body; it fetches from the cosmos what gives the plant structure. Finally, with the stinging nettle, the iron that is so connected with the incarnation of the human 'I', brings the spiritual archetype of the plant to physical expression with the help of sulphur and maintains contact with the archetypal plant in the sphere beyond the cosmos, permeating it with intelligence. With these first three preparations the soil is so far developed that all four of the members connected with the physical manifestation of plants – the physical, etheric, astral and 'I' – are being enhanced.[2]

In the second group of three – oak, dandelion and valerian – the cosmic, acidic side is emphasised. Sulphur is no longer explicitly referred to even though it creates one of the most powerful acids. This contradiction is resolved when we consider that in the case of the first group, sulphur is mentioned as the element that brings the spirit into the physical domain of nature. As sulphuric acid it no longer has that function and has become terrestrial. In the second group of three it concerns the acidic effect that has sunk into the physical.

In the case of oak bark, calcium is in the foreground. Most significant for its function, however, is its structure. Calcium in the oak bark takes the form of calcium oxalate. In the passage through the life processes, the oak creates with the bark a prematurely excreted yet complete flowering process. The calcium oxalate retains its total structure due to the nature of the oak and especially the sulphurous effect of oxalic acid. Gunter Gebhard expands on this:

> The oak as a Mars/iron being has a deep relationship
> to the 'I', to the sphere beyond the physical cosmos
> where the spiritual archetypes manifest themselves
> to the spiritual researcher. Calcium draws everything
> astral into itself. Form is possible in the physical world
> due to the astral, where forms have their origin; how
> the form appears originates in the idea that works
> through the astral. The calcium oxalate of the oak
> draws the macrocosmic into the microcosmic, and
> through the iron power of the oak the connection is
> made with the spiritual archetype. If the microcosm
> and macrocosm are in harmony with one another
> then the organism is healthy. The oak bark preparation
> maintains this harmony between the two poles
> and therefore furthers the health of the plant
> prophylactically via the soil.

It could be said that in oak bark the calcium is taken hold of from above, from the cosmic pole, which is expressed in the oxalic acid process. With dandelion, this process is enhanced in that the most complete and highly developed flower draws in the finely distributed silicic acid from the cosmos and brings it into connection with the living potassium in the milk sap. Again it is the acidic effect through which the forces of the cosmos impress themselves on the dandelion, and through it the soil.

Valerian rules the phosphorous process in the way described in Chapter 8.

The characteristic polar qualities of the six compost preparations are therefore inclined in the way they work together to create an overarching synthesis – namely the forming of a preparation of the evolving sun-filled centre between the heights and the depths. With the strengthening of this centre the soil, or diaphragm, can develop more and more of its own activity. On the alkaline side, the trio of yarrow, chamomile and stinging nettle open up the substances of the soil/diaphragm from below, while on the acidic side the trio of oak bark, dandelion and valerian open up the substances from above. Valerian takes on a special position in this context as the carrier of phosphorous in that it unlocks the extra-cosmic sphere of the spirit, the world of archetypes (see Diagram 10.1).

Gunter Gebhard explains further:

> In their chemical behaviour, alkalis are Luciferic in character while the acids have a somewhat Ahrimanic nature. Sulphur is likewise an image of Lucifer and phosphorous reflects Ahrimanic forces. We have to find a balance between these two forces. In the farm, taken as an organism in itself, this balancing middle

is delineated as the soil/diaphragm. It is now up to human beings to develop this middle realm.

The animal sheaths

The animal organs used in the preparations also form a sequence, the first three taking us from the metabolic pole of the animal towards the heart and the second three from the front of the animal, the sense pole, towards the middle. In connection with this, Rudolf Steiner describes the planetary influences on the animal in relation to the sun (see Diagram 10.2). He starts out by observing as follows:

> The animal organism lives in the whole complex of Nature's household. In form and colour and configuration, and in the structure and consistency of its substance from the front to the hinder parts, it is related to these influences. From the snout towards the heart, the Saturn, Jupiter and Mars influences are at work; in the heart itself the sun, and behind the heart, towards the tail, the Venus, Mercury and the moon influences.[3]

The higher animal differentiates itself horizontally in the direction of its backbone with the outer planets working in the nerve-sense pole and the inner planets in the metabolic pole. Both are orientated towards the heart but from opposite directions.

The sequence of inner planets – Moon, Mercury, Venus – used by Rudolf Steiner is taken from his spiritual research into old mystery wisdom. This also applies to the characterisation of their macrocosmic spheres vertically in the farm organism. Here the sequence downwards into the depths is Mars, Jupiter, Saturn and upwards into the heights is Moon, Mercury, Venus, Sun.

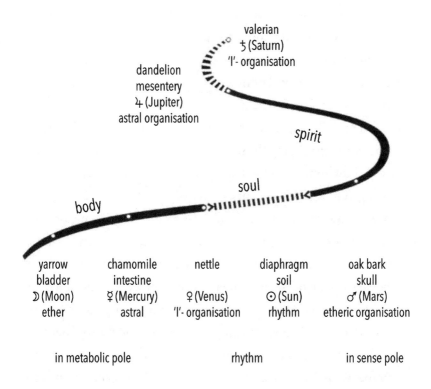

Diagram 10.2: The polar relationship of the inner and outer planets to the animal organs with regard to the development of the spiritual members of the farm organism and its threefold division into body (from below), spirit (from above) and soul (rhythmic centre).

In the animal organism, the activity of the inner planets, Moon, Mercury and Venus, is concentrated in the abdominal organs and from behind towards the heart, hence in the renal system. The kidneys take in the forces streaming in from Venus; closely connected with this are the forces of Mercury, and in the kidney-bladder system those of the moon. The Mercurial nature expresses itself in the strongly rhythmic functions of the abdominal organs. In the kidneys are both excreting and re-absorbing processes. They excrete in a mercurial way the primary urine from the arterial blood stream, sift it, readjust its substances and introduce the useful material back into the blood. This is a kind of fluid breathing process that becomes a one-sided, purely excreting process in the bladder function to expel material that is of no use to the organism. Here the moon forces are modifying those of Venus.

It is different with the taking in of food, which is sifted while being digested in the intestines. What is not usable by the organism is excreted in a more solid form. This organic process can also be seen as the expression of a Moon activity. What is fed as fluid digestive juices into the lymph and blood streams, however, can be seen as being Venus processes modified by Mercury. What takes place in the abdominal organs as Venus-renal processes in both humans and animals modified by Mercury and Moon, is found again in the activity of the stinging nettle preparation. We can say that it is here in the organic realm that the stinging nettle, together with potassium, calcium, hydrogen and iron radiations, completes the transformation of substance referred to earlier, namely the astralisation of living substance to create a new form of nitrogen out of the living plant.

The series of animal organs used for the preparations are, with one exception – the stag's bladder – taken from domestic animals (see Diagram 3.1 on page 39). In the bladder the kidney process comes to an end in that what is sifted out by the kidneys is concentrated and excreted. In the ball-like form of the bladder, the fluid element takes on the form of the mediator of Moon forces. In the process of making the preparation, it contains the sun nature of the yarrow flowers in which the earthly nature of the potassium salts are sublimated through the plant's life processes.

Deeper inside the abdomen we find the intestinal tract. At its end the large intestine microbially processes part of the remaining food residues and excretes what is of no further use. This is also a process under the influence of the moon. At its beginning is the small intestine: a long, hose-like tube in a coil with a vast internal surface area. Here the Venus activity sifts the nutrients, excreting towards the large intestine and directing the digestive juices towards the lymph and blood streams. It is supported and modified by the activity of Mercury, which brings about the rhythmic peristaltic movements of the villi, the dynamic glandular activity and the stream of substances passing through the intestinal walls. The Mercury forces stand midway between the Moon and Venus activity. They adapt themselves on the one hand to the changing circumstances and, on the other, step beyond the bounds of Mercury activity that separates inside from outside.

The chamomile flowers are stuffed into the small intestine. Their sun-like quality has transformed the earth-bound calcium and potassium into a condition that – as an extension of the

159

refreshing, vitalising effect of the yarrow preparation – helps to achieve healthy growth. That is to be understood in the sense that the archetype of the plant can bring itself to expression as a true earthly image in every stage of growth.

As the third in the group, the activity of the stinging nettle preparation reaches much closer to the centre ruled by sun forces, that is to the hidden heart-function of the soil. The astral 'inner working' of the stinging nettle is so permeated by this force of intelligence that it can do without an animal organ. It is also able to process the salt substances of potassium and calcium in its organic activity as well as transform iron into beneficial iron radiations. In this it appears to express the activity of Venus. Together with yarrow and chamomile it ensures that potassium and calcium are brought to higher levels of activity and, in addition, it calls upon the Mars character of iron to engage in radiant activity. The Venus activity, inwardly understood, creates free spaces in the living realm for new development possibilities. One could say that it doesn't nourish but makes nourishment possible. It is in this sense that the iron radiation of the stinging nettle preparation can be understood. It does not itself bring about the activity of the Venus sphere, but enables the stinging nettle as a preparation to work in the sense of regulating the iron content of the soil (or diaphragm of the farm organism) to bring healing in much the same way as occurs in the heart and circulation system.

The 'alkaline trio' makes the soil accessible to sun forces. It unlocks the forces of the earth for plant growth, and builds up the physical body of the soil-diaphragm from below. The trio of the oak bark, dandelion and valerian preparations behaves differently. Through it the realm of cosmic substance connected with the outer planetary spheres comes to fruition.

The appearance of the oak, the power in its growth, the hardness of its wood, bears the stamp of the Mars activity. The skull of the domestic animal is likewise a powerful expression of Mars activity. The brain cavity, however, is filled by a substance formed by Moon forces: the brain. In the preparation this is replaced by the Mars forces in the calcium oxalate-structured bark of the oak. The same forces that formed the skull and radiated through the moon nature of the brain, now work on the calcium of the oak bark. As oak bark preparation it works differently to the chamomile preparation as a prophylactic against external infections.

In its strongly defined form and outwardly directed sensory organs, the head has no means of extending itself with the exception – to a limited degree – of the antler-bearing animals (not those with horns). The head with its sense organs faces the world. The continuation of nerve-sense activity must be sought on a higher level. It is found among ruminant animals and developed in a remarkable way. With the sense organs of the head the cow perceives the outer world in a comparatively dull and dreamy way. With its lower senses, however, and particularly the sense of life of the peritoneum, it perceives the inner world very clearly. With the horns, whose purpose is to hold back escaping forces, it extends its intelligent nature throughout its organism. The sense activity in the head and that in the body interrelate. If the first brings sensory awareness, the latter brings supersensory strength (see Diagram 10.1 on page 150).

It is the mesentery peritoneum that perceives active spiritual forces in the bovine metabolism. The sense activity of the peritoneum, which in the cow is internally directed towards the substances, is then externally directed towards the finely distributed silicic acid in the cosmos by becoming the sheath for the dandelion flowers. It must be recognised at this point that it is the higher, etheric condition of silicic acid that is referred to. The surrounding peritoneum senses and actively draws in this silica substance and unites it with the sublimated potassium of the milk sap in the flowers. The processes occurring from below in the dandelion and from above in the peritoneum indicate an encompassing Jupiter activity. Just as all planets are the work places of hierarchical beings there are beings of wisdom on Jupiter.[4] On Jupiter wisdom is substance.[5] If the appearance and the processes active in the substances of the dandelion are brought to mind, then the wisdom-filled activity of Jupiter in this highly developed plant will become visible from the root to the leaf rosette, from the flower of shining gold to the white dandelion clock. Among flowering plants the colours of white and yellow are ascribed to Jupiter. The wisdom of Jupiter, which is reflected outwardly in the dandelion, turns inward, surrounded by the inner 'heaven' of the peritoneum, into the wisdom-filled form and function of this abdominal organ. The dandelion preparation builds a synthesis in terms of earthly form and function of this noble Jupiter activity. The powerful effect of this synthesis may be seen when the plant grows 'sensitive to all things, and will draw to itself all that it needs.'[6] We can also say that via cosmic silica the dandelion preparation gives the living wisdom of Jupiter to the organic manure, which then lives on in the plant as an evolving sensory power.

If the dandelion flowers, which are so strongly permeated by Jupiter forces, are then encased by the membrane of the mesentery, the capacity for receiving the finely distributed silicic acid from the cosmos remains potent. The fruit of this enabling process comes to the plant via the preparation and the organic manure. It works, however, in the opposite direction by making the plant sensitive to the substances it needs from the earth in order to grow. What was already indicated regarding the effect of the oak bark preparation appears in the dandelion preparation on a higher level. The living context of soil-diaphragm-plant is, so to speak, ensouled. Just as the activity of the first trio opens up the earthly substances of the root pole, the second trio opens up the non-earthly cosmic substances and forces of the flower pole.

Finally, on the highest level, we can say that the valerian preparation, via the organic manure, stimulates the phosphorous that has been freed by the decomposition processes in such a way that it can once again become the physical carrier of the intrinsically spiritual element in the living realm.

The valerian does not need to be surrounded by an animal organ, but for different reasons than the stinging nettle. It has the capacity to raise the phosphorous out of the dynamic, earthbound, living processes of the root realm, purify it in the vertical, upward-striving stem, and in the flower bring it to a condition of pure receptivity for the forces coming from the world of spiritual archetypes that find their physical-etheric carrier in phosphorous.

The nature of valerian is such that it has a strong connection to light and warmth. The composition of substances that form as the plant grows under the influence of the sun and the inner and outer planets, disintegrates when the plant dies back. Every decomposition process releases warmth. What came into existence disappears physically. Inner warmth, or the warmth ether, is also a memory in which the events and achievements of the entire living planetary system are recorded.[7] This is where the Saturn principle is active. As a cosmic memory it conserves the past, the great evolutionary connection between the coming into being of humanity and the earth and the present time.[8] This is expressed outwardly in the fact that the Saturn sphere contains and surrounds the solar system on all sides like a blanket of warmth, and wherever the valerian preparation is applied, a Saturn-like warmth blanket is created that forms a boundary between the inner and the outer.

If the valerian preparation is sprayed on to a manure or compost heap it has the effect of creating a protective sheath like the animal organs do for the other preparations. We may assume that not only are the decomposition processes limited within a boundary skin, but that the radiations from the other separately placed preparations are also held within the heap. The warmth blanket produced by the valerian preparation is woven of outer warmth (the warmth element) and inner warmth (the warmth ether). It can be thought of as an organ that receives radiations streaming in from the cosmos and in the phosphorous emerging from the saturnine process of decomposition finds a willing physical carrier.

Just as the whole round of the preparations is opened by the yarrow preparation, it is closed by the valerian preparation.

With this observation we can begin to understand that the six compost preparations are complete and a totality. They are composed in such a way that they sound together in harmony. The totality is the centre between the heights and the depths, and its coming into being is the enlivening of the earth itself.

The compost preparations working as a totality

The soil is what makes all life on the earth possible. In it are united, under the rulership of the sun, all the planetary and stellar radiations along with the forces of the earth's depths radiating up from below. This mutual interplay is the result of activity realised in past ages of the earth. It works on through the interactive wisdom in the kingdoms of nature. To recognise this unlimited, connection-filled inheritance is the basis upon which the farm is built in biodynamic agriculture. Each of the six compost preparations can make a potential contribution towards developing this 'middle realm' represented by the soil.

From this understanding come the principle methods to transform the pre-existing place, the farm, according to the laws determined by physical nature and those active in the mineral, plant and animal kingdoms. Out of this emerges, on a small scale, what the earth as a whole is – a self-contained organism. This opens the door to a path of development, enriched by the ideas of spiritual research, that raises agriculture to a higher level. In making and using the preparations, the end point becomes the beginning of a path into the future. The end points are products from the mineral, plant and animal kingdoms, the effects of the currently existing rhythms of the solar year and, finally, the conditions of the four elements of earth, water, air and warmth. Through the knowledge born of spiritual research these evolutionary end points are brought towards ideas that cannot be found in nature, and only by applying these ideas do they gradually gain a characteristic lawfulness. These ideas lead to an inversion and transformation of the past into what is of the future. The question then arises as to how, if at all, the phenomena of this inversion can be grasped by natural science. The concepts we have only relate to what we can measure, count and weigh, to what is past and not what is to come. This is where in our knowledge of nature a deep abyss opens up between the lifeless quantitative world and the qualitative phenomena of the living, ensouled and spirit-filled nature.

The 'canon' of the six compost preparations that are added to the freshly gathered organic materials on the farm, work on the living soil-plant complex to enhance development, according to many experiences and ongoing experimental results. Three evolutionary and interweaving steps can be described:

1. The compost preparations collectively impregnate the decomposing organic material of the compost or manure pile each in their own specific way. They give to the otherwise chaotic breakdown process an overarching organising principle. They guide the complex process of breaking down and correctly rebuilding substances, and help the compost or manure by unfolding its own life to contain itself

as an organism. Experience demonstrates that the decomposition processes then proceed more harmoniously. That can be shown above all in the comparatively rapid transformation of an astringent odour into a mild smell. Comparative experimental trials confirm that physiological and biological activity (for example, temperature and adsorption capacity) settle down and reach a state of healthy balance.[9]

2. The effects of the compost preparations are fully integrated within the manure piles after an appropriate period. The prepared manures have then attained a level of effectiveness in the living realm that exceeds what they would ordinarily have by nature. By spreading the manures out over the land, soil processes as a whole are given stability and brought to a higher level of fertility in harmony with the rhythm of the solar year. This has been experimentally demonstrated using many parameters, the most convincing of which has been the comparatively high humus level combined with higher microbiological activity. These resulted in a deepening of the humus layer as well as an increased population of earthworms and other soil-dwelling creatures.

3. Together the biodynamic preparations give the organic manures a greater level of vitality than is merely natural to the soil, and this flows through to the plants.

Ultimately the task of the compost preparations is, through increasing soil vitality, to enhance growth and fruit formation of the cultivated plants during the cycle of the year in their individual, site-specific conditions. In this middle realm of the soil, the overall significance of manuring for the soil comes to full realisation. It determines in its ongoing development how the crop integrates itself vertically and horizontally with root, stem, leaf, flower and fruit within this trinity of the heights, the depths and the developing middle. The plants are taken hold of by an organising power through which they work with the growing conditions of their environment.

The consequence of all this is found in the practical experiences and scientific investigations that repeatedly confirm that:

- Depending on species, the plant tends to develop closer to its ideal type.
- The root goes deeper into the soil, is more finely and more regularly ramified.
- Shoot growth goes through the stages of leaf metamorphosis in a more pronounced manner, thus creating substance in a more refined way.
- The physiological processes that form the fruiting organ, whether in root, stem, leaf, flower or the seed realm, as well as orchard fruits, come to stillness when they ripen. All these fruits become fully ripe and keep for longer.

- The composition of substances is formed and refined from the base to the flower, primarily in the realm of protein – for instance, in the relationship of pure protein to raw protein and essential oils.
- Yields tend towards a balanced optimum.

What can be immediately perceived is the pronounced taste and smell, the differentiated colour, tone and consistency, as well as the high digestibility and longer-lasting quality of the food produced. The frequently heard comment by customers is also revealing: 'We need to eat less to be satisfied.' All these things together indicate an enhanced nutritive value that works in a healthy way on the physical, psychological and spiritual development of the human being. The prepared manure thus serves the ongoing development of the earth and human beings. It gives a new and higher purpose to agricultural and horticultural work. Working with the preparations really does lead to a new sense for work and research, to a new form of art. The intention to manure in this way does not arise from an external incentive nor a sense of duty, but is an impulse that has become a concern of the heart. The more this is the case, the freer and more individually artistic the activity becomes. The significance of the biodynamic preparations for the earth and humanity can only be garnered on the path of spiritual knowledge. This is a challenge that to many may seem too demanding. If we go with it in an unprejudiced way however, it will soon become clear that the source of our intention does not lie in the world outside but purely in oneself. The results of spiritual research are this source. They shed light on what we initially have to feel our way towards in the dark. Inner certainty, however, gradually develops by applying these ideas. It is then not about instant results, but about work that is done out of love and enthusiasm for the work itself. As Steiner said:

That is the point, my dear friends – do not lose heart; know that it is not the momentary success that matters; it is the working on and on with iron perseverance.[10]

Note on Sources

The version of the Agriculture Course from which the quotes in this book are taken is the UK edition published by Rudolf Steiner Press in 2012. On a couple of occasions the American edition, published by the Biodynamic Farming and Gardening Association in 1993, is also referred to. This contains notes made by Rudolf Steiner for the course which are absent from the UK edition. Below is the list of lectures and discussions of the Agriculture Course with equivalent page numbers for both the UK and US editions.

Lecture	Page number	
	UK	US
Lecture 1 (June 7, 1924)	5	13
Lecture 2 (June 10, 1924)	17	27
Lecture 3 (June 11, 1924)	42	44
Address to Experimental Circle	57	180
Lecture 4 (June 12, 1924)	65	61
First discussion (June 12, 1924)	77	76
Lecture 5 (June 13, 1924)	87	89
Second discussion (June 13, 1924)	101	105
Lecture 6 (June 14, 1924)	107	111
Third discussion (June 14, 1924)	120	131
Lecture 7 (June 15, 1924)	125	138
Lecture 8 (June 16, 1924)	136	152
Fourth discussion (June 16, 1924)	152	171
Steiner's handwritten notes	–	191
Supplementary indications	161	242

Endnotes

Unless otherwise indicated, all references to the Agriculture Course (AC) are to the UK edition published by Rudolf Steiner Press.

1: Horn Manure (500)

1. AC. See lecture of June 10, 1924.
2. Steiner, Rudolf, *The Four Seasons and the Archangels*. See lecture of October 5, 1923.
3. The quantities recommended vary from country to country. In the UK the minimum amount recommended is 20 grams per acre.
4. Schwenk, Theodor, *Sensitive Chaos: The Creation of Flowing Forms in Water and Air.*
5. Wilkes, John, *Flowforms: The Rhythmic Power of Water.*
6. AC, discussion June 12, p. 76.

2: Horn Silica (501)

1. AC, p. 55.
2. Steiner, Rudolf, *The Four Seasons and the Archangels*. See lecture of October 12, 1923.
3. AC, p. 75.
4. Steiner, Rudolf, *From Sunspots to Strawberries: Answers to Questions*, p. 141.

3: Yarrow (502)

1. *AC*, p. 91.
2. *AC*, p. 94.
3. *AC*, p. 91.
4. *AC*, p. 93.
5. *AC*, p. 93.
6. *AC*, p. 92.
7. *AC*, p. 92.
8. *AC*, p. 39.

4: Chamomile (503)

1. *AC* p. 93f.
2. Bockemühl, J and Järvinen, K, *Extraordinary Plant Qualities for Biodynamics*.
3. *AC* (US edition), p. 225.
4. *AC*, p. 94.
5. *AC*, p. 94.
6. Verbal communication with Matthias König.
7. *AC*, p. 94.

5: Stinging Nettle (504)

1. *AC*, p. 96.
2. *AC*, p. 95.
3. *AC*, p. 98.
4. *AC*, p. 50.
5. *AC*, p. 50.
6. *AC* p. 50.
7. *AC*, p. 98.

6: Oak Bark (505)

1. *AC*, p. 97.
2. Bockemühl, Jochen, and Järvinen, Kari, *Extraordinary Plant Qualities for Biodynamics*.
3. *AC*, p. 97.
4. *AC*, p. 126.
5. *AC*, p. 105.
6. *AC*, p. 97.

7. *AC*, p. 97.
8. Stappung, Walter, *Die Düngerpräparate Rudolf Steiners – Herstellung und Anwendung* [The Fertiliser Preparations of Rudolf Steiner – Manufacture and Application], Selbstverlag, Rüfenach 2017.
9. *AC*, p. 97.
10. Ibid.

7: Dandelion (506)

1. *AC*, p. 99.
2. *AC*, p. 98.
3. *AC*, p. 99.
4. Bockemühl, Jochen, and Järvinen, Kari, *Extraordinary Plant Qualities for Biodynamics*.
5. *AC*, p. 99.
6. *AC*, p. 90.
7. *AC*, p. 99.
8. *AC*, p. 99.
9. *AC*, p. 99.
10. *AC*, p. 123.
11. *AC* (US edition), p. 225.
12. *AC*, p. 99.
13. *AC*, p. 99.
14. *AC*, p. 99.

8: Valerian (507)

1. *AC*, p. 100.
2. Ibid., p. 50.
3. Gunter Gebhard. From personal communication with the author.
4. *AC*, p. 100.
5. The ribosomes are composed of ribosomal RNA. They originate from the nucleolus in the cell nucleus and form the plasma of the organelles for creating protein (cosmologically the nucleolus is the equivalent of the earth in the cell's microcosm). The cell nucleus corresponds to the moon sphere and the cell membrane to the Saturn sphere. The mitochondria have access to their own naked DNA, propagate themselves in their own rhythm, and with

the iron-rich cytochromes have the primary task of cell breathing.
Cosmologically the mitochondria are connected with Mars.

6. Steiner, Rudolf, *Illness and Therapy: Spiritual-Scientific Aspects of Healing*, p. 69.
7. Steiner, Rudolf, *Illness and Therapy*, p. 30.
8. *AC*, p. 100.

9: Equisetum Tea

1. Steiner, Rudolf, *Occult Science*. See Chapter 4: Man and the Evolution of the World.
2. Florin, Jean-Michel, *Biodynamic Wine Growing: Understanding the Vine and Its Rhythms*, p. 156f.
3. *AC*, p. 118.
4. Stappung, Walter, *Die Düngerpräparate Rudolf Steiners – Herstellung und Anwendung* [The Fertiliser Preparations of Rudolf Steiner – Manufacture and Application].
5. Schiebe, Wolfgang, Wistinghausen, Christian von and Wistinghausen, Eckhard, *Biodynamic Preparations – Production Methods*, BDAA, UK 1999.
6. See Meyer, Ulrich, 'Optimierung der Kieselsäure-Extraktion aus *Equisetum arvense* – Ergebnisse für die alltägliche Praxis' [Optimization of silica extraction from *Equisetum arvense* – results for everyday practice] in Meyer, Ulrich and Pedersen, Peter Alsted (Hrsg.), *Anthroposophische Pharmazie* [Anthropsophic Pharmacy], Berlin 2016.
7. Steiner, Rudolf, *Occult Science*. See Chapter 4: Man and the Evolution of the World.
8. *AC*, p. 118.
9. Stappung, Walter, *Die Düngerpräparate Rudolf Steiners – Herstellung und Anwendung* [The Fertiliser Preparations of Rudolf Steiner – Manufacture and Application].

10: How the Six Compost Preparations Work Together

1. *AC*, p. 42f.
2. This and the following two quotations are from personal communication with Gunter Gebhard.
3. *AC*, p. 40.

4. See, for example, Rudolf Steiner's lecture cycles *The Spiritual Hierarchies and the Physical World* and *Spiritual Beings in the Heavenly Bodies and in the Kingdoms of Nature.*

5. Steiner, Rudolf, *Karmic Relationships, vol 7.* See lecture of June 8, 1924.

6. *AC*, p. 99.

7. *Karmic Relationships, vol 7.* See lecture of June 10, 1924.

8. See *Occult Science*, Chapter 4: Man and the Evolution of the World.

9. Koepf, Herbert, *The Biodynamic Farm: Agriculture in the Service of Earth and Humanity*, Anthroposophic Press, USA 1989.

10. *AC*, p. 61.

Bibliography

Bockemühl, J and Järvinen, K, *Extraordinary Plant Qualities for Biodynamics*, Floris Books, UK 2006.

Florin, Jean-Michel, *Biodynamic Wine Growing: Understanding the Vine and Its Rhythms*, Floris Books, UK 2019.

Masson, Pierre, *A Biodynamic Manual: Practical Instructions for Farmers and Gardeners*, Second Edition, Floris Books, UK 2014.

Schwenk, Theodor, *Sensitive Chaos: The Creation of Flowing Forms in Water and Air*, trans J Collis, Rudolf Steiner Press, UK 2008.

Steiner, Rudolf, *Agriculture: Spiritual Foundations for the Renewal of Agriculture* (CW327), Biodynamic Farming and Gardening Association, USA 1993.

—, *Agriculture Course: The Birth of the Biodynamic Method* (CW327), Rudolf Steiner Press, UK 2012.

—, *The Four Seasons and the Archangels* (CW229), Rudolf Steiner Press, UK 2008.

—, *From Sunspots to Strawberries: Answers to Questions* (CW354), Rudolf Steiner Press, UK 2002.

—, *Illness and Therapy: Spiritual-Scientific Aspects of Healing* (CW313), Rudolf Steiner Press, UK 2013.

—, *Karmic Relationships, vol 7* (CW239), Rudolf Steiner Press, UK 2002.

—, *Occult Science* (CW13), Rudolf Steiner Press, UK 1969.

—, *Spiritual Beings in the Heavenly Bodies and in the Kingdoms of Nature* (CW136), SteinerBooks, USA 2012.

—, *The Spiritual Hierarchies and the Physical World* (CW110), SteinerBooks, USA 2008.

Wilkes, John, *Flowforms: The Rhythmic Power of Water*, Floris Books, UK 2019.

Index

Floris
Books

For news on all our **latest books,**
and to receive **exclusive discounts,**
join our mailing list at:

florisbooks.co.uk

Plus subscribers get a FREE book
with every online order!

We will never pass your details to anyone else.

Printed in the USA
CPSIA information can be obtained
at www.ICGtesting.com
JSHW011521221024
72172JS00015B/131